THE SCIENTIFIC IDEAS OF

G. K. GILBERT

Mountain, water, and glacier: subjects of G. K. Gilbert's science. Mount Gilbert, rising 2,938 m (9,638 ft) above the waters of Harriman Fiord. U.S. Geological Survey aerial photograph by Austin Post.

The Scientific Ideas of G. K. Gilbert

An Assessment on the Occasion of the Centennial of the
United States Geological Survey (1879-1979)

Edited by

ELLIS L. YOCHELSON
U.S. Geological Survey
Museum of Natural History
Washington, D.C. 20560

SPECIAL PAPER
183

THE GEOLOGICAL SOCIETY OF AMERICA
P.O. Box 9140 · 3300 Penrose Place · Boulder, Colorado 80301

Copyright © 1980 by The Geological Society of America, Inc.
Copyright is not claimed on any material prepared by
Government employees within the scope of their employment.
Library of Congress Catalog Card Number 80-67676
ISBN 0-8137-2183-0

Published by
THE GEOLOGICAL SOCIETY OF AMERICA, INC.
P.O. Box 9140, 3300 Penrose Place
Boulder, Colorado 80301

Printed in the United States of America

Contents

Preface

The United States Geological Survey was established on March 3, 1879, for "the classification of the public lands, and the examination of the geological structure, mineral resources, and products of the national domain." Determining how to meaningfully note the Survey's Centennial in 1979 was difficult for a government bureau that prides itself on being a scientific establishment and tries not to overstate its contributions to modern life. Still, the Survey's Centennial was an opportunity to reflect on past accomplishments and to look forward to greater challenges and achievements in the future.

Centennial commemorations usually showcase the best. Grove Karl Gilbert was the best-known scientist on the U.S. Geological Survey's original staff. Perhaps he was well known simply because he was a pioneer in a number of fields of geology, but he is also as noteworthy in those areas in which he was not the first to break ground. Overstatements are dangerous, but Gilbert may have been the best scientist that the Survey has ever employed.

Gilbert worked hard at assigned tasks even though his particular research interests suffered. Self-sacrifice is not always a noble virtue, but the element of ruthlessness in pursuing one's own narrow objectives, which is occasionally seen today, would have distressed him. He was a loyal employee, but when the organization makes mistakes, as it certainly did, and as it certainly will, he did not follow blindly.

One reasonably well known story is that Gilbert's first scientific investigation was of a mammoth. He recoiled from the bones themselves and studied the pothole which had entrapped the beast; if there is one field of earth history on which Gilbert did not write, it is paleontology. Why then should a paleontologist be concerned with Gilbert? The short answer is that he did write on geologic time. The long answer is why not? When the U.S. Geological Survey wanted to note its centennial, someone had to get it started, and I was one of those involved, just as when the Survey wanted ground water studied, G. K. Gilbert moved into a new field. Even if the work is not in your USGS job description, if it is important, you do it.

As a member of the USGS Centennial Committee, I formally thank The Geological Society of America for publishing this special paper. However, in producing it, the Society also notes that Gilbert was the only person who twice served as president of that organization. Perhaps that says as much about his contributions to the profession of geology as anything.

By and large, Gilbert was a lone worker. Few of his works are jointly authored, in contrast to today's works where multiple authorship is common. Many years ago I found out that the best way to survive in the modern world was to team up with a smart partner. I started on this vast project with half-formed ideas, but fortunately Claude Albritton of Southern Methodist University supplied ideas, suggested potential authors, and gave thoughtful counsel. His one annoying feature is that he insists on being an anonymous partner.

The largest single collection of photographs in the U.S. Geological Survey's Field Records and Photographic Library is that of Gilbert. Nearly three thousand prints are available in his collection. Some of the authors in this book have used illustrations published earlier, but also included herein are previously unpublished photographs taken by Gilbert to help illustrate his concepts. A treasure chest—to use one more cliché—of Gilbertiana is waiting to be tapped in the Library. The key to unlock it is the knowledge of its curator, Marjorie Dalechek of the U.S. Geological Survey in Denver, Colorado.

Sooner or later a geologist is supposed to go into the field. G. K. Gilbert would have approved of this activity, but he would have been certain that all of his responsibilities were met before he left. I, instead, turned the burden of completing this volume over to Clifford Nelson of the U.S. Geological Survey, Reston, Virginia, and went to a place where no one could reach me by telephone. His help is appreciated.

Occasionally, trivia comes to the fore in the preparation of any work. How did Gilbert normally refer to himself? Grove Karl, G. K., Grove K., or G. Karl Gilbert are all possibilities. Everyone involved in the project agreed that the last two possibilities were unlikely, and eventually most authors agreed on the initials. Styles do change in manner of address as in other aspects of life. During the early days of the U.S. Geological Survey, a letter from a professional colleague would be addressed to "Dear Mr. Gilbert," whereas a friend would write "Dear Gilbert." I believe that Gilbert would have approved the use of his name in the familiar sense by the authors of this volume.

Finally, thanks to the authors. It is impossible to have a collection of essays without writers. Everyone who was asked to participate agreed to do so immediately. Everyone produced a paper despite other pressing duties. Each article considers a different aspect of Gilbert's work. These essays were written because the authors were interested in both geology and ideas. I believe that Gilbert would have approved of a critical evaluation of his work as a meaningful contribution to the U.S. Geological Survey's Centennial, for he was his own harshest critic. I need only add that the authors have written with Gilbertian clarity!

Gilbert was not a figure of myth, but a human being. A collection of letters was prepared by his friends to be presented on his 75th birthday. Unfortunately, Gilbert died shortly before the festive event. I suppose of all the tributes he might have most enjoyed was that from Henry Gannett, his long-time Survey colleague:

They may call you geologic,
Geographic, & some more,
Seismographic, philosophic,
A scientist galore—

They may talk of your acquaintance
With those antiquated lakes,
Of your knowledge and your prophesies
Of past & future quakes—

But to me is left the pleasure
Of mentioning your skill
At a little board with little holes
And little pegs to fill—

'Tis here your genius shows itself
And when you win the game
We know it's due to science
And we hoist G. K. to fame!

Good Reading!

ELLIS L. YOCHELSON

Geological Society of America
Special Paper 183
1980

'A great engine of research'— G. K. Gilbert and the U.S. Geological Survey

STEPHEN J. PYNE*

National Humanities Center, P.O. Box 12256, Research Triangle Park, North Carolina 27709

ABSTRACT

Grove Karl Gilbert served with the U.S. Geological Survey for 39 yr, from its inception in 1879 until his death in 1918. Thanks to his reputation as an explorer and to his friendship with John Wesley Powell, Gilbert occupied many administrative positions— chief of the Division of the Great Basin, chief of the Appalachian Division, chief geologist (1888–1892), and head of the section on physiographic geology. Gilbert was responsible for establishing the first hydraulic laboratory in the Survey, served on numerous committees to set cartographic nomenclature and style, supervised a series of correlation essays, represented the Survey at some international congresses, advised Powell on practically all matters relating to Survey administration, oversaw bibliographic compilations, edited manuscripts, and, in general, set an example for scientific "investigators" who did not wish to be teachers.

Identifying his career with the Survey brought both rewards and costs. Gilbert's opportunities for research dried up; nearly all studies were done on his own time and at his own expense. His major work of this period was *Lake Bonneville*, and he tried to follow it with similar works on the Great Lakes and lunar maria. But lacking time and support for full studies, he investigated instead the processes of scientific thought itself in several important methodological essays. The Survey nonetheless gave this nonacademician a responsible job, brought him into a social environment which he enjoyed, and ultimately returned him to the field at age 62 for some of his finest work, the hydraulic mining studies in California.

GILBERT AS EXPLORER

Grove Karl Gilbert began his scientific career as an explorer, ended it as a field geologist, and spent many of the intervening years as an administrator with the U.S. Geological Survey. Gilbert was a charter member of the Survey from its formation in 1879, and he identified his scientific career with it. "A great engine of research," as he once termed it, he remained with the Survey for 39 yr until his death in 1918. But between the explorer and the

administrator there existed a tension. Gilbert was always better astride a mule than behind a desk (Fig. 1A, 1B). As a result of helping to run one engine of research, the Survey, that other engine, Gilbert himself, almost ceased to run at all.

A graduate of the University of Rochester, a former clerk at Henry Ward's Cosmos Hall, and a volunteer assistant for 2 yr on the Ohio State Survey, Gilbert spent 4 yr in the Far West with the Wheeler Survey (1871–74) and another 5 yr with the Powell Survey (1875–79). It was during these years "astride the Occidental mule," as he put it, that he made his major scientific discoveries— announcing two novel forms of mountain-building (the basin range and the laccolith), developing his ideas on Lake Bonneville, and pioneering in fluvial geomorphology.

Gilbert was not involved in the political agitation which led to the consolidation of the western surveys into the U.S. Geological Survey, but he made indirect contributions. His excellent reports, for example, gave credibility to the claim that scientific research could aid in western settlement. In John Strong Newberry's influential letters to James Garfield and Abram Hewitt urging a civilian survey, it was Gilbert (who had worked under Newberry in Ohio) who was mentioned as a man "of first rate ability...inspired by true scientific enthusiasm"—precisely the sort of man the Survey would need and attract (Darrah, 1951, p. 240–241). Gilbert also labored faithfully in the field correcting errors of earlier Powell expeditions and consequently freeing the Major for his political campaigns. When Gilbert wrote Powell in 1878 that he was "anxious to hear all about the Natl. Acad. and the consolidation of the surveys," he was writing from a spike camp at Kanab, Utah (Gilbert to Powell, 10-05-78, "Letters Received and Sent," Powell Survey).

Gilbert concluded his career on the Powell Survey by editing rough notes and a penciled manuscript on the Black Hills left incomplete by the tragic death of Henry Newton. In later years, Emmanuel de Margerie wrote to Gilbert to acknowledge that "most of what may present any general interest in Newton and Jenny's description of the Black Hills of South Dakota ought to bear your signature" (de Margerie to Gilbert, Palmer Collection). It was appropriate that Gilbert should conclude his career as an explorer with a study of structure and antecedent drainage, but in a position of editor rather than investigator. Although Powell and Gilbert used radically different language during this period, the topics of structure and drainage were favorites of both. On the

*Present address: 6021 North 22nd Avenue, Phoenix, Arizona 85015.

Figure 1A. The scholar as explorer. Gilbert standing in the White Mountains, Arizona, in 1871, with a hunter and Apache scouts who were also attached to the Wheeler Survey. Photograph courtesy of National Archives.

horizon, however, loomed a position with the new U.S. Geological Survey and the writing of his *Lake Bonneville* monograph, his next major study after the Henry Mountains. The sequence and associations are suggestive. During his years as an explorer, Gilbert's creativity represented a point of acceleration and concentration that, relative to what followed, resembled the motion of a river through a narrow gorge before debouching listlessly into the delta of a great lake.

A SURVEY ADMINISTRATOR

Under the direction of Clarence King, first director of the Survey, Gilbert took charge of the Division of the Great Basin. He had no office, an uncertain budget (temporarily funded by $2,000 from King's personal accounts), and a staff only on paper. He did have a subject that both he and King were eager to pursue—the physical history of Lake Bonneville. The broad outlines of the

problem were known; Gilbert himself had discussed them in reports for the Wheeler Survey. A further preliminary study suggested to the division chief that "the magnitude of the subject had been underrated," and for several years Gilbert continued to underrate it. Nevertheless, with King's support, Gilbert moved through his field work briskly, discovering nothing startlingly new but "merely adding details of configuration and illustrations of old ideas." In 1880 he returned with a larger force which included I. C. Russell, a topographic mapping party, and two outfits—each with a "four-mule wagon, three pack mules, and a quota of saddle mules, and upon the wagons were loaded the paraphernalia of camp life." After two seasons Gilbert regarded his study of Bonneville "as mainly a first chapter of the study of the group of continental basins known as the Great Basin." He projected half a dozen investigations into themes like sedimentation processes, shore processes, and crustal deformation, each with its own "memoir" (Gilbert, 1880, p. 25-26).

The establishment of a division office, however, proved more

Figure 1B. The explorer as administrator. Gilbert in 1898, near the end of his long career as a scientific bureaucrat in the Washington office. Photograph courtesy of U.S. Geological Survey Photographic Library, portrait no. 129.

stubborn than the management of his mules. There was considerable confusion over procedures, proper vouchers, and official titles. At one point, in mild exasperation, Gilbert wrote King from Salt Lake City: "Permit me to call attention to the fact that, while I have by invoice and purchase acquired considerable office furniture, I have as yet no official authority to hire an office." He settled that problem soon enough by "hiring an entire building and with it a yard and a barn." Top priority requests followed for fasteners, vouchers, rubber bands, and "for Mr. Russell's use, a revolving chair." He noted a variety of brands on the government mules, and suggested "the propriety of the adoption by the director of a uniform pattern." By pooling their private collections of scientific books, the geologists of the Great Basin division overcame the deficits of a "frontier town." Gilbert implemented a thorough system of cataloguing "by means of cards" (Gilbert to King, 07-22-80, 11-16-80, 10-09-81, "Letters Received and Sent," U.S. Geological Survey).

All of this ended precipitately with the ascent of John Wesley

Powell to the directorship in 1881. Gilbert was promptly transferred to the Washington office, his field work on Lake Bonneville was terminated, and the publication of his monograph on it was agonizingly postponed until 1890 (see Fig. 2A, 2B). It is no coincidence that Gilbert's most prolonged period of scientific drought coincided with Powell's tenure as director. Not until the year before Powell's resignation in 1894 did Gilbert return to the field; not until Powell's death in 1902 did Gilbert recover the brilliance of his early days. Gilbert's administrative career with the Survey, both in scope and in personal costs, is understandable only by virtue of his feeling and loyalty for Powell. The friendship between the two men was as deep as either knew. But, sadly, far from liberating Gilbert's talents, association with the Major largely suppressed them. Gilbert was content to remain a silent adjutant of the man who had pulled Gilbert's career into a wide orbit around his own.

Gilbert's first chore upon joining the Powell territorial survey in 1875 was mundane enough, to arrange and classify Powell's

Figure 2A. Exploration of the Great Basin—the Stockton Bar. Plate IX from the U.S. Geological Survey Monograph 1, drawn by William Henry Holmes, slightly reduced from original size. This panorama of ancient Lake Bonneville illustrates the scene and themes of Gilbert's work on this study. First, he worked in isolation. Second, his correlations were physical rather than paleontological, extending from the horizon provided by the ancient shoreline. Third, he preferred the plane table to a rock hammer as an instrument for acquiring basic data.

miscellaneous collection of fossils at the Smithsonian Institution. From his years as a clerk at Cosmos Hall Gilbert knew the routine. Perhaps buoyed by his recent marriage and his transfer from the Wheeler Survey to Powell's, he approached the subject with an excess of gusto. The result did not amuse William Healy Dall, who wrote F. B. Meek:

We have been a good deal annoyed since I returned by the ungentlemanly behavior of a person named Gilbert—now in Powell's service. On the pretense of working on Powell's fossils he made a perfect bear garden of the room, without making any sign of apology or excuse and wanted Prof. Baird to open your library cases in order that he might use your books for study. Of course Prof. Baird declined to do any such thing and for a time at least we are rid of this nuisance. [Dall to Meek, 01-21-75, Letters Received, Meek Papers]

Dall later re-evaluated the ungentlemanly nuisance and gave a "handsome new ancestor" to Gilbert by naming a fossil after him.

Scientific and administrative housekeeping was precisely what Major Powell wanted from his adjutant in this new organization. Powell also administered the Bureau of Ethnology concurrently, and compounding these chores with agitation for land reform left little time for the particulars of running the Survey. While Powell mapped political strategies, he needed someone to arrange his administrative fossils, and increasingly this assignment fell to his loyal companion. In 1881, Gilbert dispatched I. C. Russell to Lake Lahontan in Nevada to produce a companion study to that of Bonneville. Ironically, Russell's work would see publication before Gilbert's, and a precedent was established by which Gilbert, after an initial investigation of a subject, would turn his notes over for someone else to complete. His own time was spent in revising his reports and in "various duties connected with the general work of the office." That phrase was repeated with saddening regularity over the next decade.

This is not to say that Gilbert's labors for Powell in the national office were trivial. As a result of his silent labors, Gilbert must be considered one of the founding administrators of the Federal scientific establishment. What was lamentable was that his tour of duty deprived him of research opportunities. Nearly every investigation he performed was done during off-duty hours in the evening or by taking a formal leave of absence. Almost nothing he published came off the Government presses. Some essays were simply presented orally to local scientific societies, never seeing print at all. His administrative assignment turned his famous equanimity of judgment from adjudicating geologic problems into supervising geologists. As Powell's closest adviser, he contributed immeasurably to the character and practical machinery of the new Survey, as it became, in his words, a "scientific trust." To Bailey Willis it was apparent that Gilbert was

Powell's better half. Perhaps no one else ever thought of them in that way, but in constant relations with the two I learned to know how much Gilbert, the true scientist, contributed to the geological thinking of Powell, the man of action. I do not think that they themselves were conscious of the degree to which the latter absorbed and gave out as his own the ideas that the former had silently passed through. But as Gilbert's assistant, I was sometimes jealous for that generous soul and devoted friend. [Willis, 1947, p. 33]

Even to Harvard observers like William Morris Davis it was obvious that "an important share of his [Gilbert's] thought was represented in the plans and assignments announced in administrative reports over the director's name instead of his own" (Davis, 1926, p. 120). An old companion from the Wheeler Survey, Henry Henshaw, noted that even in decisions involving the Bureau of Ethnology Powell habitually relied on Gilbert's sound judgment (Henshaw letter, Palmer Collection). In fact, as S. F. Emmons reported to George Becker, "Gilbert very particularly suggested that he (Powell) should hire another building and put his ethnologists into it, leaving this for the geologists. . . ." Nor was Gilbert completely insensitive to the quirks of his boss and comrade. As Emmons put it, "I asked Gilbert this morning what impression he got that he (Powell) would be likely to do about it. Gilbert said he will think it over for a few days, come to half a dozen minds about it, and then decide suddenly without any reference to what we had said about it" (Emmons to Becker, 04-29-82, General Correspondence, Becker Papers). Too often Powell handled his ideas as he did his fossil collections, leaving the uncomplaining Gilbert to straighten them out.

Until 1883 Gilbert continued to supervise the Great Basin studies of W J McGee and Willard Johnson, in addition to the studies of Russell. But except for a reconnaissance of Bonneville and the eastern front of the Sierras when he rendezvoused with Russell for a few weeks and inspected the fault line of the 1872 Lone Pine

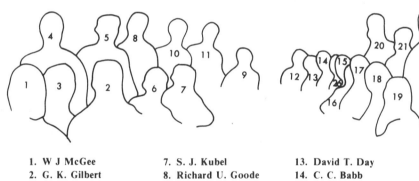

1. W J McGee	7. S. J. Kubel	13. David T. Day	19. Major J. W. Powell
2. G. K. Gilbert	8. Richard U. Goode	14. C. C. Babb	20. William A. Croffut
3. Marcus Baker	9. E. W. Parker	15. Henry Gannett	21. Robert T. Hill
4. J. S. Diller	10. H. M. Wilson	16. Wells M. Sawyer	22. De Lancey Gill
5. H. C. Rizer	11. H. W. Turner	17. F. H. Newell	23. C. Willard Hayes
6. J. C. Pilling	12. Frank Sutton	18. J. Stanley Brown	24. William H. Holmes

25. Charles D. Walcott
26. Robert H. Chapman

Figure 2B. Exploration of the Great Basin. The Old Guard passes; the retirement dinner given for John Wesley Powell in 1894 on the second floor of the Hoee Iron Building, Survey headquarters in Washington. Photograph courtesy of U.S. Geological Survey Photographic Library, portrait no. 174.

earthquake, he managed the division from Washington. The Bonneville monograph languished—eventually brought out piecemeal in the annual reports like a scientific serial—and his personal investigation of the Great Basin evaporated. Despite his devotion to Powell, Gilbert withdrew from this promising field with remorse. He had grown up on the shores of Lake Ontario, and lakes were a continual fascination to him. In his annual report for 1883–1884, he wrote the director:

While I recognize fully the considerations which led to the closing of this investigation of the Great Basin, and while the wisdom of your decision is unquestioned, I yet find myself unable to lay the work aside without the tribute of regret and expression of a hope that it may some day be resumed by another if not by myself. [Gilbert, 1883, p. 32]

He turned over his notes to his assistants, notably Russell.

In 1884 Powell transferred Gilbert to the Appalachian Division, again as geologist-in-charge. The field was new to him and the existing literature large. Gilbert's talents were maximized when he worked in original subjects or in unexplored regions, but the Appalachian assignment offered neither. He scanned the terrane for three weeks in 1884, touring the mountains of North Carolina, Tennessee, and Georgia. To supplement the field research, he initiated an extensive bibliographic index, which swelled to 11,000 cards in a year. He organized the geologic work of the division as sensibly as he could, reserving for himself a study of the relict stream terraces in the mountains. "I will personally examine the terrace system of the Atlantic drainage," he wrote Powell, "with the intention of ultimately correlating the terraces with the levelling deposits of the coast" (Gilbert to Powell, 03-30-84, Letters Received, U.S. Geological Survey). In this way Gilbert hoped to at least transfer the lessons of Lake Bonneville to its eastern equivalents—the shorelines of the Great Lakes and the Atlantic seaboard.

He never managed much time for the project, and in the end he again handed over his notes and observations to others, like F. B. Taylor. This happened also with the experimental analysis of Appalachian structure conducted in the end by Bailey Willis. Gilbert accepted the outcome philosophically: "As I do not believe in the establishment of scientific preserves I have no complaints to make and only a shade of regret that I am not in it; otherwise, I am proud of the way the work is being done" (Gilbert to Powell, 03-30-84, Letters Received, U.S. Geological Survey). Taylor and Willis were equally generous. Taylor named an arm of the Pleistocene Great Lakes that had lapped over Rochester the "Gilbert Gulf," and Willis memorialized Gilbert in the preface to his book, *Geologic Structures*. "To G. K. Gilbert, who thirty-eight years ago suggested the study of the mechanical principles underlying structural geology, I owe my introduction to the subject. 'The Mechanics of Appalachian Structure' was inspired by him."

As Powell sought him out more and more as an official confidante, and as his equable judgment was directed more and more to problems involving maps, principles of correlation, and administrative trivia, Gilbert grudgingly surrendered his field research almost entirely. His summer excursions to upstate New York resembled working vacations more than field trips. If Wheeler had bound him to the rigid time schedule of military parties dedicated to topographic mapping, dangling whole mountains before him in tantalizing frustration, the Survey under Powell's leadership disheartened him no less by its bureaucratic regimen and its own monomania for topographic maps, which compelled him to look at mountains in which he had no interest. Each month the dolorous report announced that he still worked to prepare his Bonneville report for the publication, and each month lamented that "office duties" preoccupied him.

Gilbert expended considerable time on the map question. He found that the maps were frequently inadequate as a practical tool, occasionally even "incomprehensible," and in every field study he undertook he had to conduct his own cartographic survey before beginning the specifically geologic investigation. In the *Second Annual Report*, Gilbert had closely coached Powell on announced policies of the national maps; later he chaired a Survey committee to review questions pertaining to official cartography, such as colors, nomenclature, and order of publication, and he acted as an editorial supervisor on a series of bulletins on correlation principles. After the formation of the Irrigation Survey in 1888, in Powell's haste, and against Gilbert's firm warnings, many of the maps were sloppily done; some were merely drawn from earlier hachure maps rather than resurveyed. Gilbert defended the map program before congressional committees in 1890 as a good investment, but his personal experiences belied this public vote of confidence.

These endeavors at standards taxed Gilbert's time, especially as he fought on the side of the "conservatives" to prevent a universal classification and correlation scheme from saddling geology with a ruinous tyranny of arbitrary taxonomies. A committee, he observed, "may regulate the art of the geologist, but it must not attempt to regulate his science" (Gilbert, 1887, p. 334). There were only three subjects in geology that, in his mind, required a regulated nomenclature—petrography, paleontology, and stratigraphy. These were precisely the fields Gilbert as geologist avoided. The work of the committee, which continued for several years, stood until 1902–1903 when, again with Gilbert as chairman, revisions were undertaken. But while such enterprises were useful to the Survey, they took Gilbert far from his chosen fields.

There were few breaks in the routine, all of them working vacations. In 1884 Gilbert helped Powell escort a group of scientific dignitaries from Britain on a tour of the Southwest. Having brought the insights of the West to bear on Eastern geology, in 1888 he carried them across the Atlantic to the Fourth International Geological Congress. Although he attended in a quasi-official capacity, Gilbert financed the trip entirely out of his own pocket, thus complementing his scientific field research. He had a "jolly trip," he announced when it was over, but a side excursion to Paris evoked a comment, perhaps not out of character from a man who had explored the natural marvels of the American West and who kept his financial books closely. At the sight of the Arc de Triomphe and the Cathedral of the Sacre Coeur, the Yankee, accountant, and engineer in Gilbert rebelled. He could only exclaim, "What a worthless use of money!" (Davis, 1926, p. 185).

CHIEF GEOLOGIST

When he returned, Gilbert found the Survey deeply involved in a program of land reform and agrarian conservation that centered, intellectually, on reclamation schemes and, institutionally, on the Irrigation Survey. Powell offered Gilbert the post as director of this new branch. It was an attractive invitation, for the "Western fever," as Gilbert called it, still burned bright at age 46, and the practical consequences of the program as well as its emphasis on engineering were powerful inducements. It could amount, perhaps, to a theoretical and quantitative continuation of his fluvial studies begun at the Henry Mountains. However, *Lake Bonneville* had not yet found its way to the printer, and he still yearned to expand the Great Lakes studies into an eastern equivalent of Bonneville. There were also personal reasons to decline the job, for his wife was wasting into invalidism and his adolescent sons needed more attention. The post went instead to Clarence Dutton, and Powell promoted Gilbert to the post of chief geologist. That meant, in effect, that he ran the geologic branch of the Survey, while Dutton handled the Irrigation branch, Henry Gannett the topographic, Charles D. Walcott the paleontologic, and Powell the political maneuverings of both the Survey and the Bureau of Ethnology. A year later Powell selected Gilbert to serve as chairman on the committee whose assignment was to standardize map symbols, formations, and correlation procedures. Soon thereafter Gilbert was assigned to supervise all bibliographic compilations by the Survey.

To Gilbert it must have seemed as though his career were approaching base level. He was a very competent administrator—there is universal consensus on that—but he was a greater researcher. When the Allison Commission in 1885 commended the Survey on its "business methods," some of that credit belonged to Gilbert, a self-proclaimed "stickler for accountability." But he was easier to replace as a business manager than as a field geologist. Institutionally, the Survey benefited by his promotion; scientifically, it suffered. As the years passed, the Survey's chief geologist looked more and more like its chief clerk.

His tenure as chief geologist from 1888 to 1892 was brief but distinguished. By erasing personality rather than asserting it, he reduced some of the tensions that had developed between the Powell and King cliques. He was socially a member in good standing of the Powell circle, as manifest by his charter membership in the Great Basin Mess, an informal luncheon group. But he was intellectually closer to the physicists, engineers, and chemists whom King had brought into the Survey. He was thus able to communicate with both groups. He detested controversy that crossed from science to society. Quiet but with a good sense of humor, judicious but warm, scrupulous yet considerate, Gilbert performed a frequently trivial round of editorial and administrative duties. It is no surprise to learn that he considered academic posts at Cornell and Columbia; the surprise is the strength of the friendship to Powell that prevented his accepting such offers.

Likewise, it is not surprising to learn that when the International Geological Congress convened in America in 1891, Gilbert contributed heavily to its success and helped organize the cross-continent excursions following the principles he learned from his trip to Britain. The occasion also let him exercise his quiet wit. In determining the sleeping arrangements, he decided that "we shall herd the Germans at one end and the French at the other and interpose a dining car in the middle of the train to pull them still further apart" (Davis, 1926, p. 181).

As an editor, Gilbert could be scornful or considerate. Questions of priority could be handled brusquely. To one disgruntled plaintiff, unhappy at not receiving the the recognition that he felt was due him, Gilbert wrote sharply:

In my opinion it makes little difference to the scientific world by whom discoveries are made, and I regard public discussions of questions of authorship and priority as a burden to the literature of science, occupying space and costing energy that could be better devoted. In my own writings I endeavor to give credit to those whose ideas and work I use, but I do not demand that others shall treat my work in the same way, and I do not propose ever to make reclamation of ideas borrowed or observations duplicated by others. [Davis, 1926, p. 170]

Gilbert never did. His scrupulousness, in fact, drove him in the other direction, causing him to accuse himself of "plagiarism" in instances that were clearly cases of independent discovery. He corrected his errors publicly on the grounds that it was preferable to amend himself than suffer someone else to do it for him. As chief geologist, he tried to make this a Survey practice.

Normally Gilbert edited with tact. In one instance he returned a manuscript with the comment that "the manuscript is pervaded by the originality of your amanuensis, and I fear that our editor, in eliminating that, may fail to attain that combination of accuracy and grace which would result from your own careful revision"

(Davis, 1926, p. 169). If there were complaints about his administration, they lay here—in his circumspection about making decisions that involved people. The impartiality and deliberation he brought to geologic problems occasionally aggravated personnel problems when he tried to apply the same method. He related to geologic nature more readily than to human nature. In his official conduct, as in his habits of field research, he worked better alone, but as chief geologist he did not enjoy that option.

He made a superb adjutant, but a poor surrogate. In 1889, for example, Powell was unable to attend the annual meeting of the American Association for the Advancement of Science. As retiring president, the Major had prepared an address, "The Evolution of Music from Dance to Symphony," which he asked Gilbert to deliver for him. Both in prose and speculation it was Powell the romanticist at his more effulgent. It is difficult to imagine a theme and style more foreign to Gilbert's own. Though he made a gallant attempt to present it as Powell would have wished, the result must have bordered on parody. Had he assumed the directorship of the Survey—and it was offered—something similar might have resulted. The regret in Gilbert's career is not that he failed to achieve the directorship, but that he made it as far as chief geologist. Every office like a "presidency," he once wrote, is "one of the things for which one is twice glad"—once for the honor of appointment, and again for relief upon its surrender (Davis, 1926, p. 192). Gilbert's second blessing came in 1892.

Powell's direction of the Survey, especially in matters of western land reform, had earned him enemies, many of them in Congress. When the Major succeeded in having attached to the Irrigation Act of 1888 a clause that closed the public domain until the Survey completed a survey of irrigable lands and reservoir sites, he gave his critics a common cause. He aggravated the situation by insisting that the irrigation work necessitated a national topographic map and bootlegged appropriations to achieve it. He tried to centralize western settlement just as he attempted to consolidate all scientific work of the Government under a single department. By 1890 Powell had overreached himself. Congress gutted the topographic map and irrigation projects; by 1892, the Survey itself reeled under draconian retaliations, as its budget was halved.

Gilbert had remarkably little to do with the political storm that swirled around Powell and the Survey. He left the fighting to the Major. His sole contribution was an appearance before congressional committees in 1890, in which he argued that the topographic maps were valuable for engineering and geologic work. He envisioned the Survey as a purely scientific bureau, not a political instrument. Gilbert labored dutifully in its daily administration as a research institution, but he did nothing to insure its political survival otherwise. He preferred to do himself what he would prefer the Survey to do, namely, practice science, and he painstakingly nursed *Lake Bonneville* into an almost anticlimactic publication.

Similarly in 1892, rather than lobby Congressmen, Gilbert made an unsuccessful effort to solicit funds from New England philanthropists to finance a project that became a lifelong dream with him: to drill deeply into the Earth's crust to make some crucial geophysical determinations. Meanwhile he revived field research through the only device left open to him, an official leave of absence; on leave, he studied the crater at Coon Butte east of Flagstaff, Arizona. When the political storm over Survey mapping and conservation practices struck full blast in 1892, Gilbert was supplementing his analysis of Coon Butte scenery by studying the moon in evenings at the Naval Observatory in Washington. What time he did not give to Survey administration, he gave to scientific research. Controversy, regardless of origin, only marred science. If anything, his philosophical abstraction from the practical political affairs of the Survey worked against the organization. One Congressman used the affair as a means of taking the Survey to task. "So useless has the Survey become," he huffed, "that one of its most distinguished members has no better way to employ his time than to sit up all night gaping at the moon" (Manning, 1967, p. 212–213). Gilbert bore the stigma proudly.

With appropriations slashed, the Survey restructured itself. Many positions vaporized, among them that of chief geologist. The responsibility for notifying the dismissed members fell on Gilbert. He did his best to locate positions at state surveys and universities for those whose positions were eliminated. Powell was badly strained, and his stump of an arm flared painfully. Like an ever-faithful Achates, Gilbert consoled him, and tried to lessen administrative burdens; he even foresaw some benefits to the Survey as a scientific institution. "While the Congressional onslaught is disastrous to many individuals, and therefore grieves me greatly," he explained, "it is not an unmixed blessing for the Survey" (Davis, 1926, p. 171).

Though the position of chief geologist was eliminated by congressional decree, he continued at it unofficially at Powell's behest. He was in that role when Powell resigned as director in 1894. Gilbert had come full circle in his dutiful labors for his charismatic friend. His final chores as chief geologist found him, in Powell's words, doing "some work on the collections of last summer, especially sorting out, labelling, and transmitting collections pertaining to mineralogy and ethnology" (Powell to Gilbert, 03-24-94, Letters Sent, U.S. Geological Survey).

AN ELDER STATESMAN

Despite the Survey's declining fortunes, Gilbert stayed on, refusing a university appointment. At age 19, fresh from college, he had tried to be a schoolmaster, but the youth of frontier Michigan sent the young scholar of Greek and mathematics packing before his first year was out. Even apart from that trauma, Gilbert seemed temperamentally unsuited for pedagogy; in his essays on scientific method, he carefully distinguished between the teacher and the investigator, between precept and example, and proudly listed himself among the investigators. At age 51, his mind continued to flash with insight, no duller for its relative drubbing in administrative routine. His health remained sound. "In the ten years of Western mountain work and camp life," as he notified a friend, "I built up my constitution so as to be a very vigorous man. Eight years of diminishing outdoor life and increasing desk work have lowered my tone somewhat and a sickness still more, but I am still insurable and fairly active" (Davis, 1926, p. 179).

With retrenchment of the Survey, Powell dispatched Gilbert west for the 1893 season; Gilbert took to the field again eagerly. Part of the critique of Powell's irrigation program was that it had ignored artesian wells, especially in the High Plains, so in one final attempt to rally his forces before being forced to resign, Powell had ordered his old guard to those sites. Powell's successor in 1894, Charles Doolittle Walcott, continued the program, and for three seasons Gilbert labored near Pueblo, Colorado, in what was literally an inversion of his talents. Instead of the structural variety of the basin range, he faced the monotonous High Plains. In place of great surface lakes and rivers, he pondered the subsurface

fluctuations and currents of aquifers. Instead of seeing his exceptional powers released for original investigation, he edited existing theories into a diluted form suitable for popular consumption. The wisdom of Gilbert's decision to remain with the Survey and return to field geology would not become apparent for more than another decade.

Under any pretext a tent was preferable to a desk. Gilbert glowed in the experience of "luxuriating in camp life on the Plains." He explained, "Everything but the entomology is charming" (Davis, 1926, p. 183). Gilbert mapped, surveyed, and diligently searched out "economic" topics such as building stone, fire clay, potential shale ores, and the sites of artesian wells. As official research it was not much, and he had ample time to ship off dozens of crates of fossils, archaeological curios, and even life specimens of Plains fauna to the national office. Unofficially, he found topics in isostasy and rhythmic sedimentation which are among his most provocative essays (Fig. 3A, 3B).

The Survey was being eclipsed institutionally no less than politically. As the number of geologists multiplied, geologic information grew exponentially. By 1894 the U.S. Geological Survey was too small to contain it all, with or without mauling by Congress. Gilbert recognized what the recession meant to him personally. To friends he confided that there is "no probability that I shall ever complete the Pleistocene studies I began in the Erie and Ontario basins" (Davis, 1926, p. 126). He had allied his career with that of the Survey, and the facts of the moment suggested that the two would rise or subside together.

Gilbert was off the Survey payroll as often as he was on it. For months at a time he absented himself, taking leave without pay. Some of the time he spent delivering lectures to colleges, some time went into the composition of dozens of articles for Johnson's *Universal Cyclopedia,* and some time went into field work on topics like Niagara that were of long-standing personal interest. So fragmented was his research time that the normal form of publication became the abstract and essay, rather than the monograph.

In 1899 his wife, Fannie Gilbert, died. Her health had subsided steadily for almost two decades, leaving her practically an invalid as the end approached. Gilbert's old exploring companions, Dutton and William Henry Holmes, came to the rescue. Dazzled by Dutton's erudition and charm, a San Antonio banker agreed to finance a trip to Mexico for the lot of them. "Shall be delighted to see you once more," Dutton exclaimed in his invitation, "and recall old times when we were young and beautiful and when the roses bloomed—or rather when the coyotes howled and the cactus spines got into our shins" (Dutton to Holmes, Random Records, v. 8, p. 91–94.) The excursion turned out to be much more urbane than the old days. When it concluded, Gilbert hurried off to join the Harriman Alaska Expedition, a constellation of talent that looked like the assembled membership of a western explorer's club. Gilbert assumed charge of glacial studies during the two month reconnaissance, using the opportunity to synthesize much of his thoughts on glacial physics and geomorphology, and spent a number of nights camped on glacial outwash plains with John Muir, among others.

Meanwhile, as Gilbert completed his report for the Harriman Expedition, controversy over the structural history of the basin range heated up. Gilbert's concept of normal faulting came under attack, and he took to the field to defend it. "Among my interesting finds," he wrote mockingly to a friend, "are a number of mistakes made by Gilbert, one of the Wheeler geologists, in 1872; but he was

Figure 3A. Western fever. Gilbert in 1891 at the San Francisco volcanic field near Meteor Crater, Arizona. Unable to locate any direct evidence of meteoric impact for the crater east of this locale, Gilbert suggested that volcanic magma might have reacted with ground water to create an explosion crater. Photograph courtesy of U.S. Geological Survey Photographic Library, Gilbert no. 801.

Figure 3B. Western fever. A typical Gilbert camp in the Great Basin. It could stand for the 1872 and 1873 seasons with the Wheeler Survey when he first announced his concepts of basin-range faulting and the sequence of flooding in Lake Bonneville. It could stand for the early 1880s when he toured with the U.S. Geological Survey and elaborated on these ideas, and for the early 1900s and his solitary return to defend his structural ideas against hostile critics. Photograph courtesy U.S. Geological Survey Photographic Library, Gilbert no. 1914.

substantially on the right track as regards Basin Range structure" (Davis, 1926, p. 240). He re-experienced a familiar dilemma: "the only remedy I know for a Utah desert wind is to camp in the lee of an irrigated farm, and that raises the question of preference— mosquitoes or dust" (Davis, 1926, p. 241). Trying to rush out of a cave to witness a flash flood left the old explorer with "some scars trying to lift the roof of a cave" with his head (Gilbert to Arch Gilbert, 05-13-11, Coates Collection).

The question remained whether Gilbert was still a serious student of the science or merely an emeritus explorer indulging his whims. When the accounts were totaled at the end of this second decade of Survey affiliation, they concluded on a dismal note. A few years later it was even more depressing. By 1902 his wife was dead, Powell was dead, and his brother Roy, his last tie to the family home ("the Nutshell"), was also dead. His two sons were grown to lead independent lives. Gilbert himself lived alone in a

Washington hotel. At age 60 he ebbed from the social and intellectual shores of geology. In 1900 he was awarded the Wollaston Medal of the Geological Society of London, Britain's most prestigious geological prize. He was only the third American to be so honored, but he wrote a friend only half-mockingly that the thing he "was proudest of" was the prize he had "won at progressive euchre" that week (Davis, 1926, p. 295).

His production of original scientific papers was becoming leaner. Nearly all were dependent on research conducted outside the Survey. Over the past decade Gilbert had ranged widely but inconclusively. His revived defense of the basin-range concept disappeared when a mysterious fire swept away his notes and maps. A proposal to the Carnegie Institution of Washington to sponsor a drilling project deep into the Earth's crust came to naught; after a preliminary grant to select a site and make cost estimates, the Institution decided that the price tag was prohibitive. His brilliant depiction of research methodology at Coon Butte, in which he described how he came to reject the hypothesis of a meteoritic origin for the crater, ended in a spectacular failure. The site is renowned today as the famous Meteor Crater. In a sense, that episode symbolizes both Gilbert's talent and frustration—the methodology and skill existed, but languished for lack of a suitable subject and adequate research support. For a while Gilbert even abandoned his own distinctive approach to geomorphology in favor of Davisian terminology. The fad did not last long, if it ever really took. But one might appropriately borrow the Davisian interpretation of a fundamental Gilbert invention, the graded stream, to symbolize the status of the old man at this crucial juncture in his career. Gilbert had "reached grade."

GILBERT'S SCIENCE AND THE SURVEY

It would be wrong to conclude that Gilbert's relationship to the Survey was one of duty grudgingly done and frustration unwillingly endured. The Survey brought him rewards as well. He had a permanent job, a position of responsibility with one of the premier research institutions in the world. Suited neither for academia nor for museum work, he at least was saved for science through his position with the Survey. He had companionship; always shy and a bit reserved, Gilbert welcomed the company of his fellow geologists. As his home life deteriorated, Gilbert turned even more to the scientific societies and semiofficial bodies, like the Cosmos Club and the Great Basin Mess. Compared to the period of his major monographs, the time between 1881 and 1902 was an intellectual drought. Compared to a more reasonable standard, however, those years also saw outstanding contributions from Gilbert. Many of the works were essays and remained latent with insight which Gilbert did not have the opportunity to explore. They were pure Gilbert, but incomplete Gilbert.

Gilbert's continued affiliation with the Survey extracted a personal price by burdening an outstanding researcher with administration. But when the Survey sent him to California in 1905 to investigate hydraulic mining debris, it gave Gilbert the chance to recover any lost opportunities. Gilbert entered the Survey on the basis of his reputation as an explorer; he left it with accolades as a theoretician, philosopher of method, experimentalist, and field geologist of the first rank. The *Geology of the Henry Mountains* (1877) had made his reputation in the nineteenth century and brought him into the U.S. Geological Survey. But it was his hydraulic studies in California that propelled that reputation into

the twentieth century, and it was a success he willingly shared with the Survey.

The only large scientific work that Gilbert produced during his years as a Survey administrator was *Lake Bonneville*. The research was completed during two field seasons, with the support of Director King. After Powell succeeded King, Gilbert's pace slowed to a crawl; he required 9 yr to complete the publication and released parts piecemeal with annual reports. By 1890 its ideas no less than its prose had been worked and reworked like the pebbles on the shore of Bonneville itself, and the monograph was stale, even a little passé. Nevertheless, the work was immediately proclaimed a masterpiece, and, barred from further field expeditions to the Great Basin, Gilbert promptly looked for substitutes. It is obvious that he longed to write a companion study analyzing the physical history of the Great Lakes. From childhood days in Rochester, lakes, rivers, and waterfalls had fascinated him. On the Ohio Survey, he had reconstructed the history of Erie's shoreline; on the Wheeler and Powell Surveys, he had studied the fluctuating levels of the Great Salt Lake; for the early Survey, he depicted the history of Pleistocene Lake Bonneville; and now, stationed in the East, he returned with what time he could spare to the study of the Great Lakes, the history of their outlets (especially Niagara Falls), and their isostatic adjustments. Nothing comprehensive resulted, and Gilbert was led to discover yet another surrogate, the lunar maria, which he could investigate from the Naval Observatory and his attic window.

"The Moon's Face: A Study of the Origin of Its Features," delivered in 1892 to the Philosophical Society of Washington and published in its proceedings, is premium Gilbert. Gilbert brought to the moon the geomorphological and geophysical techniques he had developed for the study of Lake Bonneville and the Great Lakes. The moonlet theory he proposed marked the advent of modern theories of lunar origin and the beginning of extraterrestrial geomorphology. Yet it is for *Lake Bonneville* that Gilbert will be most remembered, rather than the "Moon's Face." The gleam of the moonlet theory was a reflected brilliance.

Unfortunately, if the moon study was a projection of Gilbert's scientific aspirations, its own spinoff, the investigation of Coon Butte, represented their burial. Hoping to find evidence of meteoritic impact, Gilbert approached the crater in the expectation of finding a "buried star." Instead he transformed his research into a study of scientific methodology. The three papers on method that he published between 1885 and 1896 show a distressing picture of a man of exceptional research abilities who was compelled to study science rather than practice it. Just as his administrative chores gave the Survey more than he received from it, the end product of his methodological analysis may have meant more for the science as a whole than another monograph would have. Gilbert brought to the processes of scientific thought the same techniques he applied to the processes of nature. The emphasis of his methodology was the conviction that creative scientific thought was guided by analogies and proceeded by imitating examples of great scientific works, not by following formal precepts and research recipes communicated in textbooks or in classrooms. That he was compelled to lecture on the subject rather than set further examples through his own research was, to say the least, ironic. For a nonacademician who never wrote a college textbook, this situation might have had serious consequences for the dissemination of his influence and the maintenance of his reputation. By staying with the Survey, Gilbert denied himself the opportunity to establish a "school" of geologic thinking such as emerged under William

Morris Davis at Harvard, T. C. Chamberlin at Chicago, and Dana at Yale. Perhaps, in partial compensation, Gilbert corresponded spontaneously with young geologists whose work he liked, but his monographs were his real bridge to the future.

GILBERT, THE MAN AND THE MIND

Gilbert was never without friends or fans, and what the Survey under Powell may have taken away from him in terms of opportunities for research, it restored to him in the opportunity to enjoy scientific society. Gilbert's home life was not altogether fortunate. In 1883, his daughter Betsy, whom he "loved more than anything else in the world," died of diptheria. Eventually the disease struck both father and mother as well. Both recovered, but the emotional trauma of his first child's death stayed with Gilbert and prolonged his slow recovery. He retired for a month's rest in the Virginia countryside, and during this time "fought out" and dismissed any lingering religious sentiments that he had inherited from his father. His tendency toward solitude was reinforced; scientific men were practically his sole society.

Even before Betsy's death, Gilbert's wife Fannie began to lapse into chronic invalidism. Coal-gas poisoning early in 1883 sickened the family, but Fannie recovered more slowly than the others. The death of her daughter worsened her condition. Gilbert faithfully assumed the administrative duties of the house. Increasingly, his wife had to withdraw for visits to convalescent hospitals, while his two sons were sent to boarding schools, summer camps, and even the family home at Rochester. Though his thrift was legendary, the household required money, and Gilbert turned his hand to writing encyclopedia articles as a means of supplementing his office salary. Still the Gilbert house served as a frequent depot for neighborhood children in search of milk and gingersnaps, and remembering his own hardships in college, Gilbert began to lend money interest-free to needy students at Cornell. He proudly mended his own clothes, in his words, at "80% savings." This accountant-like attention to his household finances was almost identical to his approach to scientific studies. Nature, too, had to balance her books, especially her energy expenditures. Equilibrium was as fundamental to nature as a squared account to personal finances.

Those who knew Gilbert well spoke often about his "philosophy of self-control," which they greatly admired. William Morris Davis recalled,

As deeply as certain times of unhappiness were impressed upon Gilbert's self, as frankly as they were spoken of to a very few, they were never made known to the greater number of his associates and they must now lie buried with him; buried all the deeper because his courageous philosophy of life led him to live joyously. He kept his griefs and disappointments to himself and radiated only good cheer upon his comrades. [Davis, 1926, p. 9–10]

In what is probably the best description of Gilbert by a contemporary, W. C. Mendenhall wrote that

in sheer balanced mental power, Gilbert was probably unsurpassed by any geologist of his time. Fundamental among the qualities of his mind were self-knowledge and self-control. These qualities he possessed in a degree equalled by few. That mind which he knew and controlled so well was a quiet, efficient, powerful instrument, which functioned perfectly. Thus he was the very antithesis of the brilliant, temperamental, erratic genius. He recognized both his powers and his limitations, and did not undertake that which he was not equipped to do. [Mendenhall, 1920, p. 42]

Like the geologic world he described, Gilbert's personality was an equilibrium which never worked to excess, either intellectually or emotionally. He never advanced romantic theories of nature, and rarely placed himself in a position of emotional outburst. He left that to others, notably the Major. He rarely attended theaters, and eventually retired from them so much that he would buy tickets for friends to escort Fannie to a show while he stayed home. If he did attend a play, he avoided sentimental melodramas, for he could break down into weeping. The same could happen at evening readings at home among friends, a common pastime at the Gilbert household. At this point, Henry Henshaw recalled, "unable to go on," he would "temporarily relinquish the book to another" (Henshaw letter, Palmer Papers). A perennial favorite for reading was Clarence King's *Mountaineering in the Sierra Nevada*.

His companions were geologists, and his society that of scientific clubs. His favorite was the celebrated Great Basin Mess. It was organized in 1881 when the members of the lame-duck Great Basin Division (Johnson, Russell, McGee, and Gilbert) convened in a Survey room for brown-bag lunches. Occasionally they even sat on bed rolls and packing crates in imitation of a Bonneville field camp. Before long, the Mess evolved into an indoor picnic, with a different member each week packing a lunch for the whole group. Soon new members were initiated, until finally, swollen in size and ceremony, a professional caterer and hired room were required. The Great Basin Mess became a semiofficial institution of the Survey; visiting geologists dined by invitation, and the Mess was famed for its good spirit.

That the Mess had the side-effect of dividing the Survey between King's men and Powell's men was something Gilbert deplored. The function of a society, as he informed the Cosmos Club on its 25th anniversary, was "to bind the scientific men of Washington by a social tie and thus promote the solidarity which is important to their proper work and influence." He continued,

The world but imperfectly realizes that its progress in civilization is absolutely dependent on science. . . . The influence of our scientific corps— an influence of national and more than national extent—is strong in proportion as it is united and it suffers from every jealousy and needless antagonism.

With that he toasted the anniversary of the Cosmos Club as the "silver wedding of Science and Culture" (Holmes Random Records, v.1, p. 157).

Gilbert's social success is all the more remarkable when it is realized to what extent he differed intellectually from most of his companions—not that this implied a difference in the quality of his work, although a friend from his Ohio days once confided that he found "more real ideas" in the Henry Mountain and Lake Bonneville monographs "than in a cord (wood measure) of the [Survey] bulletins" (J. J. Stevenson to Gilbert 10–12–12, U.S. Geological Survey Field Records).

What really differentiated Gilbert from many of his fellow geologists was his conception of geologic time. A pure Newtonian at heart, Gilbert sought to explain geologic phenomena by bold mechanical analogies, or experiments seeking to establish mechanical models of natural processes. When he was graduated from college, he confessed to a friend that he had found engineering more attractive than geology. It was his methodological genius to bring the techniques of engineering to the subject matter of geology. He was indifferent to geologic history either as conceived by stratigraphers and paleontologists or by geophysicists. Both subscribed to historicism, to the acceptance of "time's arrow."

Figure 4A. Rivers in field. Gilbert along the Genesee River not far from his hometown of Rochester, New York. As a youth, Gilbert rowed on its waters, foreshadowing more famous scientific encounters with rivers in his later years. Photograph courtesy of U.S. Geological Survey Photographic Library, Gilbert no. 1958.

Whether measured by the fossil record or by various entropy clocks, time was progressive. But for Gilbert, geologic events represented a complex equilibrium between force and resistance. He never wrote historical geology, distrusted paleontology, ignored stratigraphy except as a source of physical measurements, and expressed deep skepticism about the methods by which geophysicists measured the age of the Earth. Nearly all of Gilbert's studies are defined by periods of equilibrium, not stages of evolution. All begin with the onset of some disequilibrating event (an intrusion of magma, the overcharging of a stream with debris, the flooding of a basin with water) and conclude when equilibrium is restored. Standard historical periods are ignored. Equally, Gilbert did not subscribe to the social evolutionism of Powell; he remained as uninterested in the evolutionary future of social institutions as in the evolutionary past of the Earth.

Perhaps because of his relative lack of research commitments, it was during these doldrum days with the Survey that he advanced with two short essays some alternatives to this conception of geologic time. One, based on the High Plains research into artesian wells, sought to find a "rhythmic" chronometer in bedding sequences, and the other, a presidential address to the AAAS in 1900, proposed an active search for other rhythmic timepieces, suggesting, as an example, the precession of the Earth.

But different as such perceptions might be from those commonly held, Gilbert never allowed them to intrude into the social environment of science. Vigorously attacked at a professional meeting on his basin-range hypothesis, for example, Gilbert politely declined to answer his challenger, thereby defusing a potential confrontation. Asked why he did not answer, Gilbert laughed and explained that had no wish to be too hard on his attacker; the facts would speak for themselves.

A SCIENTIFIC RENAISSANCE

At age 60 Gilbert seemed to have eased into a career as an elder statesman of geology. For more than a decade, out of loyalty to Powell, he had replaced major research with administrative assignments. During another decade, he had assumed, at substantical cost in time, the offices of a great many scientific societies. He appeared to have dissipated his energies and entered a long, graceful denouement—writing elegant articles for encyclopedias, interpreting the field for younger geologists, looking philosophically at the nature of science, touring as a sort of scientific celebrity with International Geological Congresses and the Harriman Expedition, perfecting a scientific method which had little substance—and he encountered frustration at nearly every attempt to initiate new research projects or revive old ones.

Though it seemed so, like the great mass beneath Lake Bonneville, depressed by the overburden of water, there remained an elasticity. Like his model for the Earth's crust, Gilbert's personality could be both locally rigid and broadly plastic. When the climate changed, when the burdens of the past evaporated, Gilbert slowly rebounded. One career was seemingly sliding to an end, but another was beginning. With Powell's death in 1902, Gilbert wrote obituaries, executed the estate, and looked after his old friend's widow. Then he looked to his own future. His final years, though broken by illness, brought perhaps his finest studies. His reddish-brown hair paled to white, but his blue eyes burned as clearly as ever.

Director Walcott turned over to the Survey's most distinguished scholar of fluvial geology a comprehensive study of the hydraulic mining question in California. Gilbert returned again to the western mountains and to a remarkable outburst of creative

Figure 4B. Rivers in flume. One of the four flumes constructed by Gilbert at the University of California, Berkeley, as part of the first U.S. Geological Survey hydraulic laboratory. From experiments, Gilbert developed a "rational," that is, a mathematical-mechanical theory for stream transport. Gilbert thus made the transition from wilderness explorer to engineer-geologist of the man-made landscape. Photograph courtesy of U.S. Geological Survey Photographic Library, Gilbert no. 3408.

energy. (Fig. 4A, 4B.) Gilbert changed his residence as well as his role. In Washington he lived with the family of C. Hart Merriam, head of the Biological Survey and a man shaped in the Powell mode; it is not too much to say, perhaps, that Merriam became a Powell surrogate. In California he divided his time between the Faculty Club at Berkeley and various field stations. Sacramento he found so dull that, as he wrote his son, he had "backslidden" to playing billiards in a public pool hall.

He joined the Sierra Club and by repeating in a recreational way the format of his exploring years, recovered his analytical powers. Essays on exfoliation, isostasy, glacial erosion, and the convexity of hilltops resulted. In 1906, when the San Francisco earthquake struck, Gilbert was in bed in Berkeley, but immediately

undertook a series of measurements—timing the intervals between waves, watching the direction of the swinging chandelier, and calculating the burning time for wooden buildings. He was appointed to both the state and Federal earthquake investigating committees. He even found time to compile years of fragmented research on Niagara Falls into a critical summary. In 1909 a stroke nearly finished him, and it was years before he recovered sufficiently to work full time.

All of these were interruptions to his central project, a laboratory and field investigation of hydraulic mining debris and its effect on the Sacramento River. The laboratory summary appeared in 1914 as a professional paper, "The Transportation of Debris by Running Water." The most comprehensive study of its kind in

English, it was based on the first hydraulic laboratory of the Survey and the first flume in the United States to investigate geologic questions, rather than simply the movement of water. The field report, "Hydraulic Mining Debris in the Sierra Nevada," followed in 1917; it is the foundation for the management of the Sacramento River, a masterpiece of geology and engineering, and perhaps the first geologic study to address the synthetic landscape rather primitive nature. In his flumes Gilbert scrutinized the boundary layer between water and sand, but in his field research it was the more turbulent boundary between man and nature that he analyzed. The opportunity to study the hydraulic mining question, he noted in his preface, was a chance to revive studies of fluvial geology which he had first begun at the Henry Mountains before he joined the Survey. Now, at the end of his Survey career, he had come full circle. He revived and exhaustively described his old subjects. Even more impressively, he made the transition from wilderness explorer to engineer-geologist of the synthetic landscape; he studied an engineered environment as much as a natural one. His geomorphology did not pertain to slow infinitesimal processes acting over eons of geologic time, but to almost catastrophic interruptions of natural equilibria created by human manipulation of the landscape.

When the second report appeared, Gilbert was 74 yr old. Yet with scarcely a pause to look over the galley sheets, he took to the field again. This time it was to another topic of his exploring youth, the basin range. He planned to remarry, too. "Alice and I have been lovers for years," he wrote his son Arch. But until his health fully recovered from apoplexy, the elder Gilbert refused to consider marriage with Alice Eastwood, a California botanist. Now they would go ahead and together they would raise his grandson, Palmer Grove, partly orphaned by the death of Arch's wife. Plans were set for marriage after his 75th birthday.

But while visiting his sister in Jackson, Michigan, Gilbert's health collapsed. From his sick bed he labored over his accounts, demanding of himself what he had also demanded of nature, that its books balance. It required considerable effort, but he persevered, and in the end the books were settled—not only his account with nature, but his account with the Survey. The Survey had taken, but it had also given. The hydraulic mining studies not only balanced out the old themes that he had reluctantly abandoned during his years as an administrator, but they balanced out his career as a scientist as well. Gilbert had done it all—explorer, field geologist, theoretician, experimentalist, philosopher of method, engineer, administrator, founder of scientific institutions, and the only man to be elected twice to the presidency of the Geological Society of America. Now the accounts squared, and he drew a small box in his pocket notebook to indicate that fact at the bottom of the page.

A few days later he packed to leave the hospital. There were more studies beckoning in the Utah desert; with Alice Eastwood, there was a new wife; with his grandson, a new family. But his old life had run its course and the half-completed basin-range monograph would not be published for another decade. Five days before his 75th birthday, the life of this great engine of research came to an end. But the life of that other great engine of research, the U.S. Geological Survey, with which Gilbert's career was so intimately allied and which owed so much of its success to Gilbert's dedication, in 1979 began its second century.

REFERENCES CITED

Darrah, William Culp, 1951, Powell of the Colorado: Princeton, Princeton University Press.

Davis, William Morris, 1926, Biographical memoir of Grove Karl Gilbert, 1843–1918: National Academy of Sciences Biographical Memoir XXI.

Gilbert, Grove Karl, 1880, Annual Report 1: U.S. Geological Survey.

——1883, Annual Report 4, U.S. Geological Survey.

——1887, The work of the International Congress of Geologists: American Journal of Science, ser. 3, v. 34, p. 434-440.

Manning, Thomas, 1967, Government in science: The U.S. Geological Survey: Lexington, Kentucky, University of Kentucky Press.

Mendenhall William C., 1920, Memorial of Grove Karl Gilbert: Geological Society of America Bulletin, v. 31, p. 20-42.

Willis, Bailey, 1947, A Yanqui in Patagonia: Palo Alto, California, Stanford University Press.

NOTES ON UNPUBLISHED REFERENCES

George Becker Papers, Library of Congress.

Coates Collection, private collection of Gilbert letters in possession of Donald Coates, Binghamton, New York.

William Holmes Random Records, National Portrait Gallery Archives.

F. B. Meek Papers, Museum of Natural History Library.

Palmer Collection, private collection of Gilbert papers in possession of Mrs. Dorothy Palmer, Cloverdale, Oregon.

Powell Survey, Letters Received and Sent, Record Group 57, National Archives.

Pyne, Stephen J., 1976, Grove Karl Gilbert: A biography of American geology [Ph.D. thesis]: Austin, University of Texas, 658 p. Includes complete inventory of primary materials pertaining to Gilbert's career and bibliography of Gilbert's publications.

——1980, Grove Karl Gilbert [Ph. D. thesis]: Austin, Texas, University of Texas Press.

U.S. Geological Survey, Letters Received and Sent, Record Group 57, National Archives.

MANUSCRIPT RECEIVED BY THE SOCIETY SEPTEMBER 1979
MANUSCRIPT ACCEPTED MAY 20, 1980

Geological Society of America
Special Paper 183
1980

Contributions of Grove Karl Gilbert
to glacial geology east of the Mississippi River

GEORGE W. WHITE

Geology Department, 254 Natural History Building, University of Illinois, Urbana, Illinois 61801

ABSTRACT

Grove Karl Gilbert (1843–1918) began his work in professional geology in 1869 as "local assistant" on the newly established Second Geological Survey of Ohio (the "Newberry Survey"). He worked without salary but received $50 per month for expenses. Gilbert investigated Williams, Fulton, and Lucas Counties, the three most northwestern counties adjacent to the Michigan State line. The area extended from Indiana on the west to Toledo at the western end of Lake Erie. His originality, analytical power, and clear verbal and graphic exposition, which were the distinguishing characteristics of his work throughout his career, are completely exhibited in this his earliest work, published in nine articles and reports from 1871 to 1874.

Gilbert's maps of glacial geology of Williams and Fulton Counties are noteworthy for their detail and accuracy. He was the first to discover and map the major end moraines of the Maumee basin, from east to west (youngest to oldest): the Defiance, Fort Wayne, and Wabash Moraines. Gilbert found that the drift sheets of the moraines were actually multiple and that the uppermost drift was draped over a core of older material, a concept not to be widely recognized until almost 50 years later! Gilbert discovered the continuity and mapped the beach ridges now called, from highest to lowest, Maumee, Whittlesey, and Warren.

After distinguished work in the West from 1871 to 1881, Gilbert returned to the East and, among other projects, began to study the glacial geology of New York, especially Niagara Falls and the Lake Ontario raised beaches. In his classic study of the origin and retreat of Niagara Falls, he concluded that the Horseshoe Falls had retreated at the rate of 5 ± 1 ft/yr.

Gilbert's other work in the Ontario basin was on the uplifted shorelines of Lake Ontario, the character and amount of their later tilting, and the surface features and drainage channels of the Rochester region. He was impressed by the boulder pavements in the till and by the evidence of glacial and postglacial folding and faulting, on which he published several papers.

INTRODUCTION

G. K. Gilbert (1843–1918) was graduated from Rochester University in 1862, and from 1863 to 1869 was a member of the staff of Ward's Cosmos Hall (now Ward's Natural Science Establishment) in Rochester. Here Gilbert attained a wide experience in biological and geological materials and a wide knowledge of fossils, rocks, minerals, and Paleozoic stratigraphy and Pleistocene features and materials. In 1867 and 1868 he was in Albany to superintend the mounting of the Cohoes mastodon.

GILBERT'S WORK IN NORTHWESTERN OHIO

The Second Geological Survey of Ohio was organized in 1869 with John Strong Newberry as State Geologist (Merrill, 1920, p. 406; Hansen and Collins, 1979, p. 8). Newberry rapidly assembled a very capable staff of men who already were, or were destined to become, some of America's outstanding geologists (Merrill, 1920, p. 406). Gilbert applied for a position on the Ohio Survey, but Newberry could only offer the 26-yr-old Gilbert an appointment as "local assistant" to begin July 1, 1869, at $50 per month for expenses. Gilbert was assigned Williams, Fulton, and Lucas Counties along the Michigan border in northwestern Ohio, an area that extended from the Indiana line to Toledo at the western end of Lake Erie. He spent the summers of 1869 and 1970 in the field and in the winter aided Newberry in New York City, where Newberry had retained his position as professor at Columbia. Gilbert, an expert draftsman, aided in drawing fossils and helped in writing reports.

A preliminary report on Williams, Fulton, and Lucas Counties (Toledo) soon appeared (Gilbert 1871a). The same year, Gilbert (1871b) published a paper on the surface geology of the Maumee Valley. This paper was an early version of his longer paper that included the same map on a larger scale (Gilbert, 1873a). Gilbert also became acquainted with the geology, especially the glacial geology, of the area for some distance south of his assigned counties. N. H. Winchell, four years older than Gilbert, had been assigned the 16 counties of northwestern Ohio south of the three counties assigned to Gilbert. The young geologists were in communication and compared their findings and explanations. Gilbert's detailed reports and maps of each county were probably completed by 1871, and the county reports appeared in 1873 (Gilbert, 1873b, 1873c). Although he had no topographic maps, his glacial maps of Williams and Fulton Counties (Figs. 1, 2) are excellent and accurate. They bear very favorable comparison with

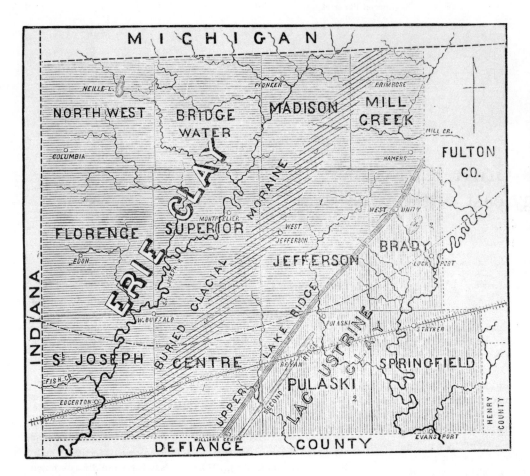

Figure 1. Map of glacial geology of Williams County, Ohio (Gilbert, 1873b). The "Buried Glacial Moraine" is the Fort Wayne Moraine of present-day usage (Goldthwait and others, 1961). Actually, this moraine is not buried, but Gilbert interpreted the material of the upper part as lacustrine rather than fine-grained till which it actually is (see Fig. 4). The upper lake ridge is the highest Maumee beach, and the second ridge is a lower Maumee beach.

the maps of Forsyth (1959, 1972) and with recent surficial deposits maps of King (1977) and Reimann (1979). Gilbert is recognized as the first to publish "detailed description and unequivocal interpretation of either terminal or recessional moraines [of continental glaciers]" (Gregory, 1918, p. 121).

In extreme northwestern Ohio, Gilbert was led to the location of moraines by analysis of the curious drainage patterns of the St. Joseph and the St. Marys Rivers. Gilbert recorded in his diary under November 10, 1870, "invented moraine hypothesis" for these stream patterns (Davis, 1927, p. 12). He asserted that the pattern of the moraines showed the lobate margin of the ice sheet in its later stages, when a lake was impounded between the waning ice sheet and the margin of the Maumee basin. He showed that the earliest lake, now called Lake Maumee, had drained to the west past Fort Wayne to the Wabash River. He next had to find some way of holding in the lake from the lower territory to the east. He soon arrived at the conclusion that the ice sheet itself had formed the dam and had gradually retreated out of the basin to the northeast.

The sandy ridges of early lake shores of a higher Lake Erie had been noted as early as 1812 (Brown, 1814, p. 147), and raised beaches of early Lake Ontario had been seen as early as 1743 by William Bartram, Lewis Evans, and Conrad Weiser (Bartram, 1751, p. 53). However, Gilbert was the first to trace the higher shore lines around the whole west end of the Erie Basin and to show the orderly sequence of at least four successive lower lake levels. Even more important, he showed that the highest lake (Lake Maumee) drained west past Fort Wayne, Indiana, to the Wabash River. These old beaches are well shown on his county maps (Figs. 1, 2)

and on his map "Raised Beaches North of the Maumee River" (Fig. 3).

As there were no topographic maps, Gilbert could only use railroad surveys. Fortunately, the Lake Shore and Michigan Southern Rail Road (now Penn-Central) extended in a straight line west-southwest across Gilbert's three counties, and Gilbert used the precise levels to construct a profile and cross-section (Fig. 3). The profile plainly shows the beaches and the Fort Wayne Moraine and Defiance Moraine. The Defiance Moraine is concealed by later lake sediments, as discovered by Gilbert. The Fort Wayne Moraine is not buried by lake sediment as Figure 3 might imply, but has a core of coarser till beneath a covering of fine till, as has been determined in great detail at Fort Wayne, Indiana, by Bleuer (1974).

Gilbert (1873a) summarized the glacial geology of the Maumee Valley not only in "his own" three counties but also in the adjacent territory in Michigan and Indiana and to the south in Ohio. By text and map (Fig. 4), he set forth clearly the drainage control by end moraines. He showed the precise location of the Fort Wayne Moraine in his area and his important discovery of the location and character of the buried Defiance Moraine. He knew the location of the Wabash Moraine between the St. Marys and Wabash Rivers, but as this was in Winchell's area he courteously left it out. He showed the two most prominent raised beaches, the Maumee and Whittlesey, and traced these throughout the basin.

Gilbert interpreted the fine-grained clay till west of the lower region of glacial lake clay as a lake deposit in a still higher and much more extensive ancient lake. Pebbles and occasional cobbles

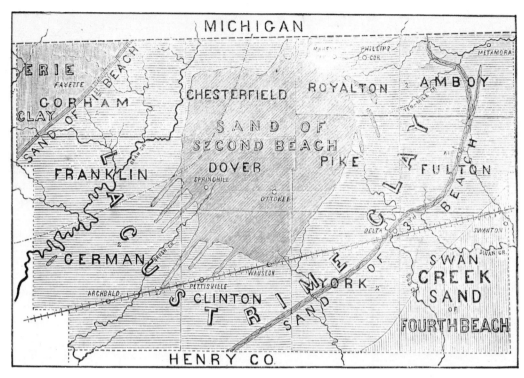

Figure 2. Map of glacial geology of Fulton County, Ohio (Gilbert, 1893c). The sand of 1st beach is upper Maumee beach. The sand of 2nd beach is also Maumee. Sand of 3rd beach is Whittlesey beach. Sand of 4th beach is Warren beach. Note that the Defiance Moraine, the existence of which Gilbert discovered, is not indicated; it is at the position of the sand of 2nd beach which overlies the lacustrine clay above the buried Defiance Moraine (see Figs. 3, 4).

and boulders were interpreted as deposits from icebergs floating in a large lake, that he believed extended very far to the south. He did interpret the cores of the moraine as truly ice-deposited material. In this iceberg origin he was following Newberry, who, along with Gilbert, gave up the iceberg origin of upper till a few years later (White, 1976, p. 53, 54).

GILBERT AND MASTODONS

Gilbert (1871c) saw and may have participated in the excavation of a mastodon, presumably in 1870, found in a peat and marl deposit at St. Johns, southeastern Auglaize County, Ohio. As this is 65 mi south of Gilbert's area, his attention must have been directed to it by Winchell. Knowing of Gilbert's experience at Ward's and his participation in the excavation of a mastodon in New York (Gilbert, 1867) and in mounting the Cohoes mastodon at Albany (Gilbert, 1871d), Winchell immediately thought of Gilbert.

According to Gilbert (1871c, p. 220), there was a question as to whether the remains "belonged strictly to the period of the deposits in which they were found. . . ," the animal having become mired and died in place, or "were the bones transported. . .into the bogs." In a closely reasoned analysis, Gilbert was able to show that the St. Johns mastodon had mired in to the knees and had expired in mud 6 ft deep, and that 2 ft of mud had accumulated since. "The [postglacial] date of the burial of the mastodon remains is as definitely recorded as that of the ice period." Gilbert (1871d) had earlier come to a similar conclusion about the Cohoes mastodon.

INTERLUDE IN THE WEST

Through Newberry's recommendation, Gilbert was appointed to the staff of Lt. G. M. Wheeler's survey in 1871 and transferred to

J. W. Powell's survey in 1874. Gilbert continued his association with Powell in the U.S. Geological Survey when it was established in 1879. His connection with Powell was to continue until Powell's retirement from the Survey in 1894. Gilbert continued with the Survey until his death in 1918. Gilbert's work in the West, which resulted in the famous Henry Mountains report and the Lake Bonneville monograph, are dealt with elsewhere in this volume. However, it should be mentioned that Gilbert's early studies of the raised beaches of Lake Ontario and Lake Erie and his knowledge of present Great Lakes shore features and their origin were very useful to him in his work in the Lake Bonneville basin in Utah and adjacent areas. The general features of existing and raised ("fossil") lake shores are dealt with in the Lake Bonneville monograph (Gilbert, 1890a) and even more extensively in a special paper (Gilbert, 1885). In his discussion and his illustrations, Gilbert's acquaintance with the shores of all the Great Lakes is apparent.

LAKE ONTARIO BASIN

In 1881 Gilbert moved from the West to Washington, D.C., and became Chief Geologist in 1889. He was able to spend part of each summer from 1885 to 1891 in study of the glacial geology of the Lake Ontario basin and adjacent regions. He had not lost interest in his earliest work in northwestern Ohio and was now able to compare the raised beaches of the different early Great Lakes with those of Lake Bonneville that he had studied so effectively and had reported in his famous monograph. Ice advance had obviously controlled the origin and expansion of the early lakes and produced a sequence of drainage outlets. Gilbert's first work in 1869 and 1870 showed that the earliest lake in the Erie basin (Lake Maumee) had drained west past Fort Wayne to the Wabash (Fig. 4). It soon became obvious to him that to the east this lake must have been held in by ice. As the ice retreated to the east, the earliest lake expanded and an eastern outlet must have come into existence

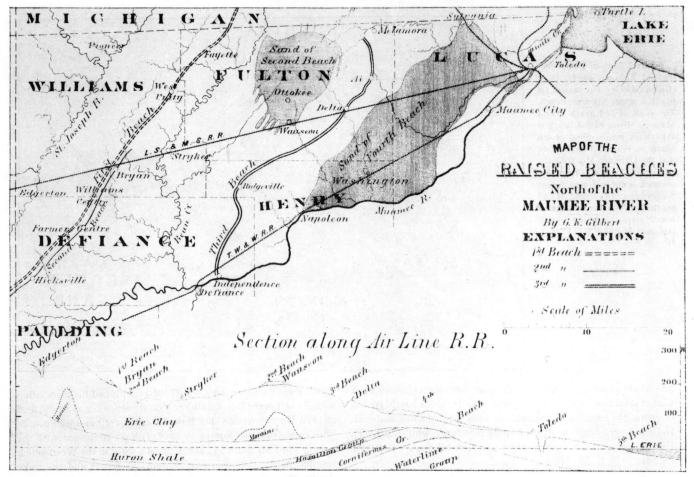

Figure 3. Map of raised beaches north of the Maumee River (Gilbert, 1973a, p. 549). The Air Line R.R. is the L.S. & M.S.R.R. (now Penn-Central). The most western buried moraine is the Fort Wayne, and the eastern buried moraine is the Defiance Moraine. (For current names of the beaches, see Figs. 1, 2.)

while ice still filled the St. Lawrence Valley. Gilbert's brilliant deduction led to finding in New York a series of eastern outlets and some of the waterfalls, now "fossil," which they had formed (Gilbert, 1897a).

When an eastern outlet was opened, the level of the large lake was lowered 550 ft, to a level below the Niagara Escarpment which had to this time been submerged. Lake Iroquois came into existence in the Ontario Basin, the water from the Erie basin flowed over the Niagara Escarpment, and the Niagara River was born. The sequence of events was described in detail in a very successful lecture at the Toronto meeting of the American Association for Advancement of Science in 1889 and published the next year (Gilbert, 1890b). The eight maps and diagrams of this paper include the frequently copied famous "section of Niagara Falls, showing the arrangement of the hard and soft strata, and illustrating a theory of the process of erosion" (Gilbert 1890b, Pl. 8), which appears in textbooks to the present day (usually copied from the Niagara Falls folio, Kindle and Taylor, 1913, Fig. 11). Gilbert's maps of the early lakes stages are among the very earliest (actually the earliest?) to show the location of these lakes (Gilbert 1890b). Gilbert was an expert draftsman and it is possible and even probable that Gilbert drafted these figures himself.

In his famous "Rate of Recession of Niagara Falls," Gilbert

(1907) demonstrated that the rate of recession varied. The different widths and depths along the 7-mi gorge were correlated with different rates of discharge as lake areas varied with opening and closing of certain outlets. He hesitated to ascribe a figure for rate of recession throughout the 7 mi. However, he did feel sure the rate of recession from 1842 to 1905 "is found to be 5 feet per annum, with an uncertainty of 1 foot" (Gilbert 1907, p. 25). The data were presented in the famous plate showing changing crest positions in historic time (Fig. 5).

About the same time, the large classic book on Niagara Falls by the Canadian geologist J. W. W. Spencer (1907) also appeared. Gilbert's lengthy review (1908) of Spencer's book was generally complimentary, but it did take issue with the calculations for age of the falls. Gilbert's disagreement was based on considerations of head, discharge, volumes, efficiency, and energy, all impressively couched in hydraulic engineering and mathematical terms, as noted in some detail by Pyne (1975). It is interesting to compare the work of Spencer (1907) in the Ontario basin with that of Gilbert. A useful paper would result from a detailed study of the interrelation of the important work of these two men whose methods were so similar in some ways and different in others.

Gilbert's mapping of the Lake Ontario raised beaches (1898a) convinced him that the beaches of Lake Iroquois had been uplifted

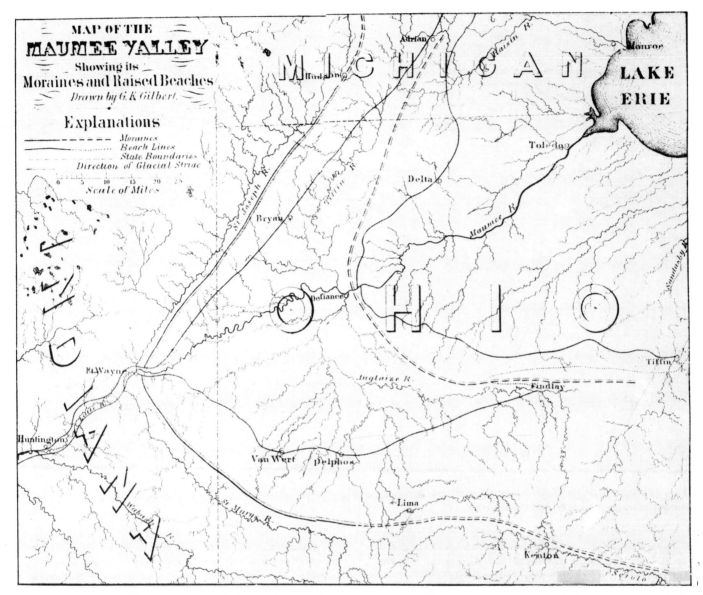

Figure 4. Map of the Maumee Valley showing its moraines and raised beaches (Gilbert, 1873a, p. 541). This is the first map to show moraines and raised beaches for this part of Ohio, Indiana, and Michigan.

350 ft to the north-northeast. By extrapolating his map of isobases (1898a, Fig. 94, p. 604) to the west, a tilt in the same direction (but not necessarily at the same amount) should be present (see also Gilbert 1897b, Fig. 7, p. 358). Later-date detailed investigations confirm this tilting for the western end of Lake Erie from Cleveland to New York State (Totten, 1977, 1979). Gilbert (1898a) made use of precise measurements, over an extended period of years, to determine recent lake level changes. His early mathematical training and his continued fondness for sophisticated mathematical approaches are exhibited in this study, as well as in many others.

The differences in lake levels and variation of discharge to the east and to the west affected the discharge at Niagara Falls and accounted for the varying widths of the gorge and the depth of the channels as Gilbert so elegantly demonstrated (1898a, p. 607).

STRUCTURAL FEATURES ASSOCIATED WITH GLACIAL DRIFT

Study and analysis of "minor features" associated with glacial deposits have assumed increasing importance, and Gilbert's early notice of them deserves emphasis in this review.

Gilbert (1898b) discovered and illustrated boulder pavements at Wilson Hill, New York. He used the position of the boulders to estimate the direction of ice motion and as an indication of two episodes of ice advance. This was no casual note, for he cited other published references on the subject and his own "unpublished observations" on other boulder pavements. He appears to have recognized that boulder pavements are far more common than generally realized and may offer serious problems in excavations, especially in tunneling (White, 1974, p. 336).

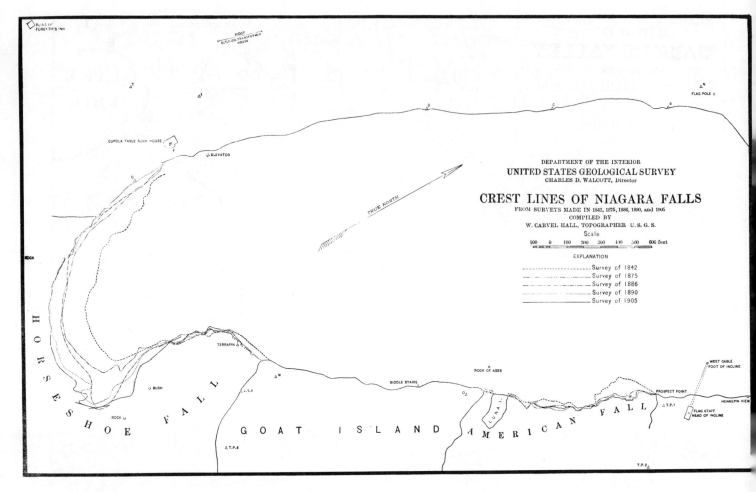

Figure 5. Crest lines of Niagara Falls (Gilbert, 1907). This was compiled by W. Carvel Hall under Gilbert's direction. This map is the basis of Gilbert's estimate of erosion rates from 1842 to the present.

Other kinds of structural features associated with the drift are small folds and faults. In "Some New Geological Wrinkles," Gilbert (1886) reported on "several small post-glacial anticlinals in the horizontal limestones of Jefferson Co., N.Y. and in the shales near Dunkirk in the western part of the State." He ascribed these to "expansion caused by the warming up of the surface layers of the rocks as they recovered from the cold of the Glacial period."

Gilbert reported on the "Postglacial anticlinal ridges near Ripley and Caledonia, New York" at the 1891 American Association for Advancement of Science Meeting in Washington. A small anticline raised the surface "six or eight feet high" and extended down at least 40 ft where exposed on the Lake Erie cliff in extreme western New York. Like other small ridges of Devonian shale in north-western Ohio and of the Trenton limestone in northern New York, Gilbert showed these to have been formed after the departure of the last ice sheet; he attributed these to the postglacial rise of temperature and consequent expansion of the rocks (Gilbert, 1891). He ascribed small anticlinal ridges in "Corniferous" lime-stone in Genessee County, New York, to sinking of blocks due to solution at depth of salt and gypsum. This is certainly an early recognition of solution at depth reflected in surface features, as noted by Wallach and Prucha (1979). The existence of such features due to solution of salt at depth is now widely known and taken into account in engineering works (Christiansen 1967, 1970).

Somewhat later, Gilbert (1899) described and illustrated an anticline passing into a thrust fault at Thirtymile Point, New York, as shown on the shore of Lake Ontario (Figs. 6, 7). He interpreted this structure as formed by ice shove just as the ice ceased to move, thus explaining the preservation of the protruding fold. The statement that this structure differed from the "ordinary anticlinal ridges of the old lake bottoms, which are clearly subsequent not only to the ice-sheet but to the glacial lakes" is tantalizing indeed. Can this be interpreted to mean that Gilbert saw so many small anticlines that form small surface ridges on the lake plain that further discussion is unnecessary? Dislocation of strata, especially Devonian shales, now near or at the surface of the uplifted lake

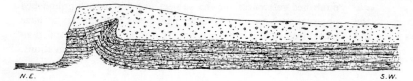

Figure 6. Section at Thirtymile Point (Gilbert 1899, Fig. 2). Note surface expression of fold.

Figure 7. Anticline at Thirtymile Point, New York (Gilbert, 1899, Pl. 12). Compare sketch in Figure 6.

bottom are indeed common, but their expression as surface ridges deserves detailed study.

The early observations of Gilbert and of others (Read, 1873; Decker, 1920; Chadwick, 1920) on relatively minor dislocations of strata and overlying drift are now becoming important not only for engineering works but also for sophisticated tectonic studies (Sbar and Sykes, 1973, and extensive references therein). Decker (1920, p. 69–81) extended Gilbert's idea on the age, especially the postglacial age of the structures, and summarized possible stresses that might account for their formation. A more recent and more sophisticated study by geotechnical engineers and engineering geologists (White and others, 1973) shed additional important light on the question of origin and time. The origin of these features has been uncertain, but the general opinion (Read, 1873; Gilbert, 1898a) has been that they are of glacial origin. However, some, but by no means all, of the dislocations affect the present surface and

show that the dislocations are of no great age. Some are known to have taken place in historical times (White and others, 1973).

The recent paper of Sbar and Sykes (1973) would have delighted Gilbert for its seismic, mathematical, and geological analyses of these features. These authors assign an important place for origin to plate-tectonics stresses and ascribe many to recent movements. Although Gilbert favored a late glacial origin, he did consider that the release of "compressive strains" rather than ice advance might have played a part in the origin of the features. It is probable that many were indeed formed by Pleistocene ice shove but others were formed much later, as indicated by surface expression or by faulted earlier till sheets truncated by an overlying horizontal till sheet (White and Totten, 1979). As preliminary site investigations by geologists and engineers are now required for location of large power plants, dams, and other large engineering works, it is necessary to accumulate and analyze all the earlier reports. Here is another instance of the value of Gilbert's keen observation, fine description, and balanced analysis of cause, and by implication, the age of the dislocations. Gilbert's carefully recorded observations are often times so applicable to present-day problems that they could well be included along with other data in the preparation of engineering geology reports or for presentation at public hearings.

CONCLUSION

Gilbert's very earliest geological work was on the glacial geology of northwestern Ohio. From the beginning his work was that of a mature scientist. He was the first to map end moraines (Fort Wayne and Defiance) and relate them to ice positions and lake-level control. He was the first to recognize that the moraines were composed of more than one layer of drift. He was also first to recognize and map the continuity of the raised beaches (Maumee, Whittlesey, and Warren) in the Maumee basin.

Gilbert's later work in the Ontario basin is distinguished by accurate studies of the higher lake levels, the sequence of drainage outlets, and the complex history of the Niagara River and Niagara Falls. The important but hitherto little-appreciated reports on "minor" structural features in the drift are now beginning to be recognized.

Gilbert, whose work was widely acclaimed in his own lifetime, is still recognized, appreciated, and admired as one of the U.S. Geological Survey's greatest geologists and one of America's greatest scientists.

ACKNOWLEDGMENTS

I am indebted to M. C. Hansen, P. F. Karrow, R. A. Struble, and M. K. White for critical reviews. Their comments and suggestions have been helpful. The help of Harriet Wallace, Librarian, University of Illinois Geology Library, in bibliographic matters is gratefully acknowledged. E. L. Yochelson provided copies, through the U.S. Geological Survey photographic laboratory, of Gilbert's figures reproduced here.

REFERENCES CITED

Bartram, John, 1751, Observations on the inhabitants, climate, soil...made by Mr. John Bartram in his travels from Pensilvania [sic] to...Lake Ontario...: London, J. Whiston and B. White.

Bleuer, N. K., 1974, Buried till ridges in the Fort Wayne, Indiana, area and their regional significance: Geological Society of America Bulletin, v. 87, p. 917–920.

Brown, S. R., 1814, Views of the campaigns of the North-Western Army &. view of the lake coast from Sandusky to Detroit: Troy, Francis Adincourt, 156 p.

Chadwick, G. H., 1920, Large fault in western New York: Geological Society of America Bulletin, v. 31, p. 117–120.

Christiansen, E. A., 1967, Collapse structures near Saskatoon, Saskatchewan, Canada: Canadian Journal of Earth Sciences, v. 4, p. 757–767.

——ed., 1970, Physical environment of Saskatoon, Canada: National Research Council of Canada Publication No. 11378, large folio, 68 p.

Davis, W. M., 1927, Biographical memoir Grove Karl Gilbert 1843–1918: National Academy of Sciences, v. 21, Fifth Memoir, 303 p.

Decker, C. E., 1920, Studies in minor folds: Chicago, University of Chicago Press, 89 p.

Forsyth, J. L., 1959, The beach ridges of northern Ohio: Ohio Geological Survey Information Circular 25, 10 p. and maps.

——1972, Glacial map of a part of northwestern Ohio, in Johnson, G. H., and Keller, J. J., Geological map of the 1° X 2° Fort Wayne Quadrangle, Indiana, Michigan, and Ohio....: Indiana Department of Natural Resources, Regional Geologic Map no. 8.

Gilbert, G. K., 1867, The American mastodon: Rochester, New York, Moore's Rural New Yorker, March 2.

——1871a, Report on the geology of Williams, Fulton, and Lucas Counties: Ohio Geological Survey, Report of Progress, 1870, pt. 7, p. 485–499.

——1871b, On certain glacial and postglacial phenomena of the Maumee Valley: American Journal of Science, 3rd ser., v. 1, p. 338–345.

——1871c, On the remains of a Mastodon from St. Johns, Auglaise County, Ohio: Lyceum Natural History New York Proceedings, v. 1, p. 220–221.

——1871d, Notes of investigations at Cohoes, with references to the circumstances of the deposition of the skeleton of the mastodon, ... under direction of James Hall: Twenty-first Annual Report of the New York State Museum of Natural History, p. 129–148.

——1873a, Surface geology of the Maumee Valley: Ohio Geological Survey, v. 1, pt. 1, p. 535–556.

——1873b, Geology of Williams County: Ohio Geological Survey, v. 1, pt. 1, p. 557–566.

——1873c, Geology of Fulton County: Ohio Geological Survey, v. 1, pt. 1, p. 567–572.

——1885, The topographic features of lake shores: U.S. Geological Survey, Fifth Annual Report, p. 69–123.

——1886, Some new geological wrinkles: American Journal of Science, v. 32, p. 324.

——1890a, Lake Bonneville: U.S. Geological Survey Monograph 1, 438 p.

——1890b, The history of Niagara River: Smithsonian Institution Forty-fifth Annual Report, 1890, p. 231–257.

——1891, Postglacial anticlinal ridges near Ripley, New York and near Caledonia, New York [abs.]: American Geologist, v. 8, p. 230–231; also Proceedings of the American Association for the Advancement of Science, v. 40, p. 249–250.

——1897a, Old tracks of Erian drainage in western New York [abs.]: Journal of Geology, v. 5, p. 109–110.

——1897b, Modification of the Great Lakes by earth movement: National Geographic Magazine, v. 8, p. 233–247; also Smithsonian Institution, Annual Report for 1898, p. 349–361, 1900.

——1898a, Recent earth movement in the Great Lakes region: U.S. Geological Survey, Eighteenth Annual Report, pt. 2, p. 601–647.

——1898b, Boulder pavement at Wilson, New York: Journal of Geology, v. 6, p. 771–775.

——1899, Dislocation at Thirtymile Point, New York: Geological Society of America Bulletin, v. 10, p. 131–134.

——1907, Rate of recession of Niagara Falls: U.S. Geological Survey Bulletin 306, p. 1–25.

——1908, Evolution of Niagara Falls, review of The falls of Niagara by Spencer, J. W. W.: Science, n.s. 28, p. 26–54.

Goldthwait, R. P., White G. W., and Forsyth, J. L., 1961, Glacial map of Ohio: U.S. Geological Survey Miscellaneous Geologic Investigations Map I-316, scale 1:500,000.

Gregory, H. E., 1918, A century of geology—steps of progress in interpretation of land forms: American Journal of Science, v. 196, p. 104–132.

Hansen, M. C., and Collins, H. R., 1979, A brief history of the Ohio Geological Survey: Ohio Journal of Science, v. 79, p. 3–14.

Kindle, E. M., and Taylor, F. B., 1913, Description of the Niagara folio:

U.S. Geological Survey Geological Atlas, Niagara Folio no. 190, 25 p. and maps.

King, J. M., 1977, Ground-water resources of Williams County, Ohio [M.S. thesis]: University of Toledo.

Merrill, G. P., 1920, Contributions to a history of American State Geological and Natural History Surveys: Smithsonian Institution, U.S. National Museum Bulletin, 109, 549 p.

Pyne, Stephen, 1975, The mind of Grove Karl Gilbert, in Melhorn, W. N., and Flemal, R. C., eds., Theories of landform development: Binghamton, State University of New York, Publications in Geomorphology, p. 277–298.

Read, M. C., 1873, Report on the geology of Ashtabula, Trumbull, Lake, and Geauga Counties: Ohio Geological Survey, v. 1, pt. 1, p. 481–533.

Reimann, M. C., 1979, Ground-water resources of Fulton County, Ohio [M.S. thesis]: University of Toledo.

Sbar, M. L., and Sykes, L. R., 1973, Contemporary compressive stress and seismicity in eastern North America; an example of intro-plate tectonics: Geological Society of America Bulletin, v. 84, p. 1861–1882.

Spencer, J. W. W., 1907, The falls of Niagara, their evolution and varying relations to the Great Lakes...: Geological Survey of Canada, 490 p.

Totten, S. M., 1977, Sangamonian and Wisconsinan strandlines along the south shore of Lake Erie, northeastern Ohio, and northwestern Pennsylvania: Geological Society of America Abstracts with Programs v. 9, no. 5, p. 659–660.

——1979, Beaches and strandlines bordering Lake Erie, in White, G. W., and Totten, S. W., Glacial geology of Ashtabula County, Ohio: Ohio Geological Survey Report of Investigation No. 112.

Wallach, J. L., and Prucha, J. J., 1979, Origin of steeply inclined fractures in central and western New York: Geological Society of America Bulletin, Part I, v. 90, p. 417–421; Part II, v. 90, p. 789–827.

White, G. W., 1974, Buried glacial geomorphology, in Coates, D. F., ed., Binghamton State University of New York, Publications in Geomorphology, p. 331–349.

——1976, Glacial geology in Ohio in 1874—the end of an era: Ohio Journal of Science, v. 76, p. 51–56.

White, G. W., and Totten, S. M., 1979, Glacial geology of Ashtabula County, Ohio: Ohio Geological Survey Report of Investigation No. 112.

White, O. L., Karrow, P. F., and MacDonald, J. R., 1973, Residual stress relief phenomena in southern Ontario: Montreal, Proceedings of 9th Canadian Rock Mechanics Symposium, p. 323–348.

MANUSCRIPT RECEIVED BY THE SOCIETY AUGUST 31, 1979
MANUSCRIPT ACCEPTED MAY 20, 1980

Geological Society of America
Special Paper 183
1980

G. K. Gilbert, on laccoliths and intrusive structures

CHARLES B. HUNT

Visiting Scholar, University of Utah, 2131 Condie Drive, Salt Lake City, Utah 84119

ABSTRACT

G. K. Gilbert's report on the Henry Mountains is classic for its contribution to knowledge of igneous structures, especially laccoliths, for its contributions to the understanding of geomorphic processes, and as an example of excellent technique in geologic reporting and writing. Present-day Ph.D. candidates and many of their faculty would do well to adopt Gilbert's technique.

His accomplishments are especially impressive when viewed in the perspective of the status of geologic knowledge at the time he did his work and the hazards accompanying his field work. His conclusions, seemingly elementary today, were received with skepticism by many of his contemporaries and were not fully accepted for about a quarter of a century. Modern surveys have confirmed his principal conclusions.

INTRODUCTION

Gilbert's report on the Henry Mountains, Utah (1877b), prepared for the Powell Survey, is classic in four ways: (1) it introduced the concept of laccoliths; (2) more importantly, it was the first report to demonstrate clearly that intrusive igneous masses can deform the rocks into which they are intruded; (3) it introduced major geomorphic concepts that are the subject of other papers in this book; and (4) it was in large part written in the field where geologic scenery inspires writing. The report sets an example of a reporting technique that could be a model for present-day geologists (Hunt, 1959).

Whole paragraphs of Gilbert's report were taken verbatim from his field notes. The notebooks include sketches of the landscapes as seen from each station. (Fig. 1). With the notebooks and the sketches in hand, it was easy to follow his route through the mountains 60 yr later when it was my good fortune to be able to resurvey the Henry Mountains (Hunt and others, 1953).

GILBERT'S TRIP

The Henry Mountains were still remote at the time of our resurvey in the 1930s, and, in those pre-jeep days, there were no roads and few trails. The area was accessible only by packtrain (Hunt, 1977, p. 95-104). Sixty years earlier when Gilbert visited the area, it was much less accessible (Fig. 2). He wrote (1877b, p. 1):

At the time of their discovery by Professor Powell, the mountains were the center of the largest unexplored district in the territory of the United States—a district which by its peculiar ruggedness had turned aside all previous travelers. . . . The cañons of the Colorado Basin. . .were by common consent avoided. . . .

Gilbert spent a week in the mountains in 1876 and returned for two months in 1877 when he did his principal work. He not only had long distances to travel by horseback, with his supplies carried by pack animals, but he was traveling beyond the settlements and in Indian country. He was well aware of the dangers. Only a few years earlier three members of Powell's party who left the group in the Grand Canyon and attempted to return to the settlements on foot were murdered by Shivwitz Indians. Soon after that incident, three members of Wheeler's party with whom Gilbert had worked were killed when Mohave Apache Indians ambushed their stagecoach near Wickenburg, Arizona. In 1875 a party of the Hayden Survey, ambushed by Indians south of the La Sal Mountains, lost their pack train and had to hike 100 mi to La Plata, Colorado. With such background of contemporary experiences, one may imagine Gilbert's inner thoughts when, on November 1, 1876, while traversing the east base of the Henry Mountains, he came on Indian tracks, and wrote:

We find today a trail made by one horse shod or partly shod, other horses barefooted, barefooted mules and barefooted colts, in all about 15 animals. There is a moccasin track with them. They came down Crescent Creek, started up the trail towards Trochus Butte and stopped; one went ahead and turned back and then all went down Crescent Creek Cañon. After an interval they returned and went up the creek again. The coming tracks were made in wet sand, the going in dry. Neither have been rained on. The tracks are much scattered.

Our last storm was October 20. From all this we infer that a party of Indians not familiar with the country came down Crescent Creek October 21 or 22 and after an interval of some days (long enough to go to the Colorado and back) returned. They were less numerous than their (15) animals. It is not unlikely that they were Navajos who had stolen stock from a stock range and were trying to cross the Colorado without going through the settlement.

That night, instead of camping in Crescent Creek (head of North Wash), Gilbert and his party moved out of the canyon and camped in a less exposed location back of Trochus Butte. On November 8 he again wrote:

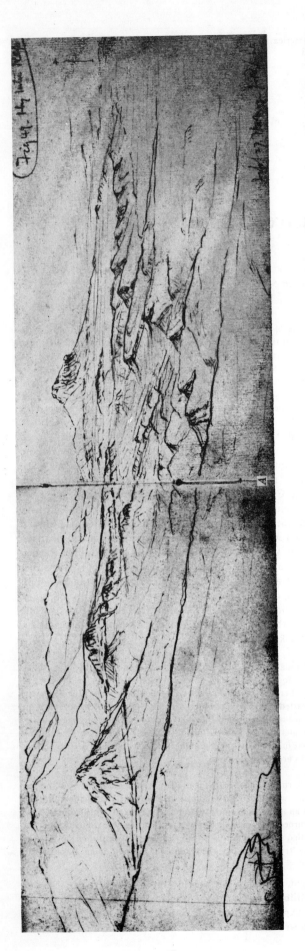

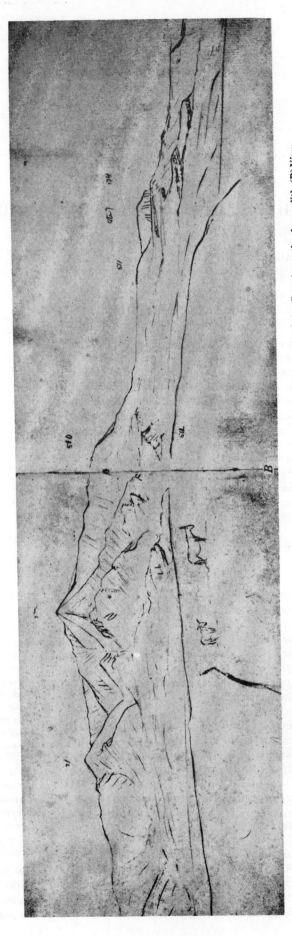

Figure 1. Sketches of the Henry Mountains, reproduced from the field notebooks of G. K. Gilbert. (A) View north along the east side of Mount Ellen. The stock is at the south end of the high part of the mountain. The conical butte at the left is Ragged Mountain, a bysmalith (Scrope laccolith of Gilbert). To the right (north) of Ragged Mountain is the Copper Ridge laccolith (Peale laccolith of Gilbert) with well-exposed floor. The butte right of center is Bull Mountain (Jukes Butte), another bysmalith. (B) View west to Mount Pennell. The ridge marked "Sta. 7" is the Horn laccolith (Sentinel Butte of Gilbert); "511" is the Dark Canyon laccolith. The Mount Pennell stock forms the peak and extends southward to Straight Creek which emerges from the mountain on the north side of station 71. Left of station 71 is Bulldog Ridge.

Yesterday we crossed the Indian trail twice. They returned westward towards Hilloid Butte (Table Mountain) with 26 animals and at least 6 pairs of moccasins. They passed eastward in two parties (one earlier than the other) crossing Cache Creek (Sweetwater Creek) near the South Twin (South Caineville Mesa).

Although Gilbert made detailed lists of his equipment and supplies, nowhere in his Henry Mountains or Lake Bonneville notes does he mention having a gun. Was the gun taken for granted, or did he travel unarmed?

Like all explorers of his day, Gilbert was accustomed to following trails of deer and Indians. Difficulties of travel, even by packtrain, were correspondingly great, and his notes showed this. He wrote on September 4:

On the march the gray mule Louisa rolls down hill with her pack a distance of 50 or 75 feet. The chief damage seems to be a cut and bruise on the thigh and another back of the ear.

September 9:

A chapter of accidents. Frank kicked by Little Nephi in the shin. Lightfoot about played out and down twice. My pack bucked off and three alfogas torn. Evening spent in repairs. Water in pockets bad.

September 10:

We have to leave first the horse Lightfoot behind and then the Baldface mule. The latter is brought in this P.M. The horse is to be sought in the morning.

Such incidents became commonplace, and on November 13 Gilbert recorded:

Panguich rolled over today into Curtis Creek. This is her third roll on the trip. Beck has accomplished two and Gomas, Joel, and Lousey one each. Our little train of 9 animals has attained to seven (eight?) rolling scrapes.

Traveling in the Henry Mountains still was an experience to be remembered 60 yr later. Toward the end of the resurvey in the 1930s, the mapping was inspected by the Branch Chief, the Chief Geologist, and Dr. Herbert Gregory. Within 4 days, each of the bosses had experienced horsefalls, two with dislocated shoulders or broken ribs, and the inspection was curtailed. During the field work, our packtrain animals fell off the trails too, and it seemed to be invariably the one packing the eggs and wine.

Gilbert was not the first to visit the Henry Mountains, and he recorded (1877b, p. 66) the following observations by others who preceded him but who did not publish their observations:

While Professor Powell's boat party was exploring the cañons of the Colorado, Mr. John F. Steward, a geologist with the party, climbed the cliff near the mouth of the Dirty Devil River and approached the eastern base of the mountains. He reported that the strata had in the mountains a quaquaversal dip, rising upon the flanks from all sides.

Steward may have had a view like that illustrated in Figure 3, but from a lower angle.

The following year Prof. A. H. Thompson then as now in charge of the geographic work of Professor Powell's survey, crossed the mountains by Penellen Pass and ascended some of the principal peaks. He noted the uprising of the strata about the bases and the presence of igneous rocks.

In 1873 Mr. E. E. Howell at that time the geologist of a division of the

Figure 2. This drawing in Gilbert's notebook is entitled "Lazarus, Duke of York." The head was reproduced in the first edition of his report with the title "Ways and Means." It was omitted from the second edition.

Wheeler Survey travelled within twelve miles of the western base of the mountains and observed the uprising of the strata.

GILBERT'S LACCOLITHS AND DEFORMATION BY INTRUSIONS

As already noted, Gilbert's report constituted a major breakthrough in understanding intrusive structures, although he generally is remembered because of his concept of laccoliths (Fig. 4). In his field notes, he first referred to the laccoliths as "bulges" and "arches," but near the middle of his trip, after examining Table Mountain (Fig. 5), he inferred their mushroom form and used the term "lacune." After his return to Washington, during the preparation of his report, the intrusions were referred to as "laculites" (1877a). In his final report, this term was modified to "laccolite."

Dana (1880) suggested that the term be changed to "laccolith" because the ending "ite" was generally used for designating kinds of rocks. This usage was generally adopted after 1900, although Gilbert evidently preferred his original term and used it as late as 1896 (Gilbert, 1896).

Gilbert's evidence that the igneous rocks in the Henry Mountains are intrusive and that they deformed their host rocks probably is repeated in many Geology I courses given today, but probably not many of today's instructors realize that they are

Figure 3. Oblique aerial view southwest to Mount Holmes (right of center) and Mount Ellsworth (left) showing the Mesozoic sandstones rising onto the mountain uplifts. Waterpocket fold in distance is formed by the same sandstones. Gilbert (1877b, p. 118–119) contrasted the summits of Mount Ellen (see Fig. 5) with those of Mount Holmes and Mount Ellsworth. He wrote that they "are not to be scaled by horses. The mountaineer must climb to reach their summits, and for part of the way use hands as well as feet." (U.S. Geological Survey Photo GS-D-26, by Fairchild Aerial Surveys, 1937)

reciting evidence from a century-old report. He wrote (1877b, p. 51–53):

What evidence. . .is there that the origin of the laccolite was subsequent to the formation of the inclosing strata rather than contemporaneous with it? May it not have been buried instead of intruded? May not the successive sheet and masses of trachyte have been spread or heaped by eruption upon the bottoms of Mesozoic seas, and successively covered by the accumulating sediments?

He found the question easy to answer.

1st. No fragment of the trachyte has been discovered in the associated strata. The constitution of the several members of the Mesozoic system in the Henry Mountains region does not differ from the general constitution of the same members elsewhere. This evidence is of a negative character, but if there were no other it would be sufficient. . . .

2d. The trachyte is in no case vesicular, and in no case fragmental. If it had been extruded on dry land or in shallow water, where the pressure upon it was not sufficient to prevent the dilation of its gases, it would have been more or less inflated, after the manner of recent lavas. If it had issued at the bottom of an ocean, the rapidity of cooling would have cracked the surface

of the flow while the interior was yet molten and in motion, and breccias of trachyte *debris* would have resulted. . . .

3d. The inclination of the arched strata proves that they have been disturbed. If the laccolites were formed in each case before the sediments which cover them, the strata must have been deposited with substantially the dips which they now possess. This is incredible. The steepest declivity of earth-slopes upon the land is 34° from the horizontal, and they have not been found to equal this under water. . . .But the strata which cover the laccolites dip in many places 45° to 60°, and in the revetments of the south base of Mount Hillers they attain 80°.

4th. It occasionally happens that a sheet, which for a certain distance has continued between two strata, breaks through one of them and strikes across the bedding to some new horizon, resuming its course between other strata. Every such sheet is unquestionably *subsequent* to the bedding.

5th. The strata which overlie as well as those which underlie laccolites and sheets, are metamorphosed in the vicinity of the trachyte, and the greatest alteration is found in the strata which are in direct contact with it. The alteration of superior strata has the same character as the alteration of inferior. This could never be the case if the trap masses were contemporaneous with the sediments. . . .

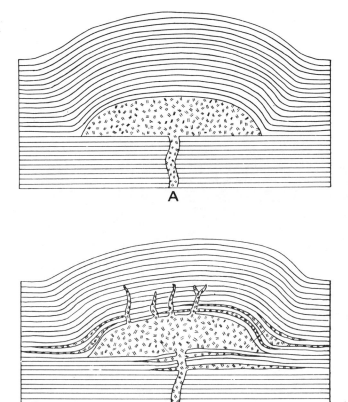

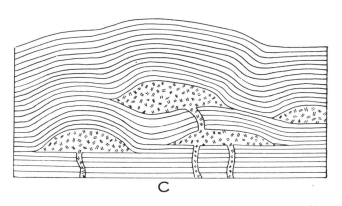

Figure 4. Gilbert's concept of the ideal form of laccoliths.

In fine, all the phenomena of the mountains are phenomena of intrusion. There is no evidence whatever of extrusion.

These are remarkable observations considering the status of knowledge about intrusions at the time. Even the igneous origin of intrusive rocks had been debated as late as the 1830s. Some recognized the significance of dikes and sills, and Murchison (1839, p. 110) introduced the term "boss" for the irregular knoblike igneous masses in the sedimentary formations of Shropshire, England. But geologists were reluctant to accept the fact that igneous masses may intrude and be younger than their host rocks. Many still followed von Buch (1836) in regarding volcanic cones as "craters of elevation", that is, the quaquaversal dips of lavas at the cones were due to doming.

Today a Geology I student would be flunked for not recognizing the significance of the evidence Gilbert cited. Yet at the time geologists were slow in accepting that evidence. Twelve years after Gilbert's report, Reyer (1888, p. 135) still argued that the laccoliths must be surface eruptions that had been buried by the overlying sediments. Neumayr (1887, p. 180) agreed that the evidence for intrusion was very convincing, but was so surprising that further confirmation was needed. Other reviews of Gilbert's report were given by Green (1879) and Davis (1924, 1925, 1926).

While Gilbert was in the Henry Mountains, his contemporaries on the Fortieth Parallel Survey were interpreting the stocks in the Wasatch Mountains, at Bingham Canyon, and elsewhere in western Utah as protuberances of Precambrian granite overlapped by the sedimentary formations around them (Hague and Emmons, 1877, p. 353). They were so considered by some until 1900, when Emmons, who had favored the Precambrian interpretation, finally accepted the evidence that the stocks in the Salt Lake City area are intrusive (Emmons, 1903).

Gilbert was puzzled as to why the laccoliths spread between the sedimentary formations and lifted the overlying rocks rather than simply continuing their route upward and erupting as volcanoes. Today, a century later, geologists still are puzzled. The problem as discussed by him was not the source and propelling force of the magma but the circumstances that determined their stopping place. He reasoned (1877b, p. 75):

The coincidence of the laccolitic structure with a certain type of igneous rock is so persistent that we cannot doubt that the rock contained in itself a condition which determined its behavior.

We are then led to conclude that the conditions which determine the results of igneous activity were the relative densities of the intruding lavas and of the invaded strata; and that the fulfillment of the general law of hydrostatics was not materially modified by the rigidity and cohesion of the strata.

His presentation of this thesis was followed by an analysis of the densities of the intrusive rocks and of the sedimentary rocks, factors influencing arching of the overlying strata, thickness of overburden, and limiting size of the mushroom shape he postulated.

Gilbert's first premise is sound, because bulging laccoliths require viscous rather than fluid melts. But few geologists accept the premise that the magma was perfectly fluid and the country rock was nearly homogeneous, as would have been required for the intrusions to conform to the laws of hydrostatics (see, however, Gussow, 1962, and Johnson and Pollard, 1973). Other equally important or more important factors were overly minimized, such as mode of emplacement of the stocks, not then recognized; discontinuities in the invaded strata; regional and local structure; composition and viscosity of the magmas; rates of intrusion; overburden as affecting domal curvature; and possible eruption.

RESURVEY OF THE HENRY MOUNTAINS

The resurvey of the geology of the Henry Mountains in more detail (Hunt and others, 1953) has been dubbed "the report that in 5 years proved Gilbert was right in two months." I agree with the assessment. The resurvey indeed proved Gilbert completely right in showing that intrusions do deform the formations into which they are intruded. The resurvey did develop minor modifications of Gilbert's concept of laccoliths.

Figure 5. Oblique aerial view of the north side of Mount Ellen. Table Mountain (lower right) suggested the mushroom shape of a laccolith. It is capped by sandstone (light patches) of the Jurassic Morrison Formation. Along the eroded edge (bottom of picture) Cretaceous Mancos Shale is exposed dragged up vertically against the faulted distal edge of the intrusion. An anticline with 500 ft of structural relief extends southward from Table Mountain toward the stock (see Fig. 6). The three ridges extending north from the summit of Mount Ellen are examples of laccoliths "jutting forth like dormer windows." as Gilbert described. South of Mount Ellen can be seen Mount Pennell (with sharp peak), Mount Hillers, and Mount Holmes. (U.S. Geological Survey Photo GS-D4, by Fairchild Aerial Surveys, 1937)

It was found that the large mountain domes (Fig. 3) were caused by intrusion of cross-cutting stocks (Figs. 6, 7), an interpretation based on surface structural geology that later was confirmed by aeromagnetic surveys. The resurvey also found that the laccoliths were satellites injected laterally from the stocks (Fig. 8). The determinations accord very well with Holmes's (1877) interpretation of the structure at Ute Mountain in southwest Colorado (Fig. 9), an interpretation made at the same time Gilbert was in the Henry Mountains, but Holmes did not provide such complete description of the evidence.

Actually, Gilbert even anticipated the shapes of the laccoliths as determined by our plane table survey by observing that some of the laccoliths "jut forth. . .like so many dormer windows" (Gilbert, 1877b, p. 38). This feature is illustrated in Figure 5.

The resurvey confirmed that the laccoliths had lifted their roofs by folding, although it was found that the folds are anticlines with axes radiating from the stocks, rather than circular domes (Fig. 8). It added the detail that some of the laccoliths, including the one at

Table Mountain (Fig. 5), had lifted their roofs by faulting, a structural form later referred to as a bysmalith (Iddings, 1898). Table Mountain has a trap-door structure with the crescentic fault on the distal side of the uplift, which is structurally open toward the stock (Fig. 6).

Probably the best known contribution of the resurvey is that unique intrusive form defined as a cactolith (Hunt and others, 1953, p. 151).

Gilbert did mistake one intrusion, the Ragged Mountain bysmalith (Scrope Butte in his report) as having an exposed floor (Figs. 1A, 6) (Gilbert, 1877b, p. 53). He stood at the well-exposed floor contact of the Copper Ridge laccolith (Figs. 1A, 6; his Peale laccolith) and looking south could see the horizontal sandstone beds around the base of Ragged Mountain, and assumed they extended under the igneous rock of the butte. But those sandstones do not extend under the intrusion; they are turned up almost vertically in a narrow zone at the faulted, distal side of the intrusion, a trap-door structure like that at Table Mountain. When

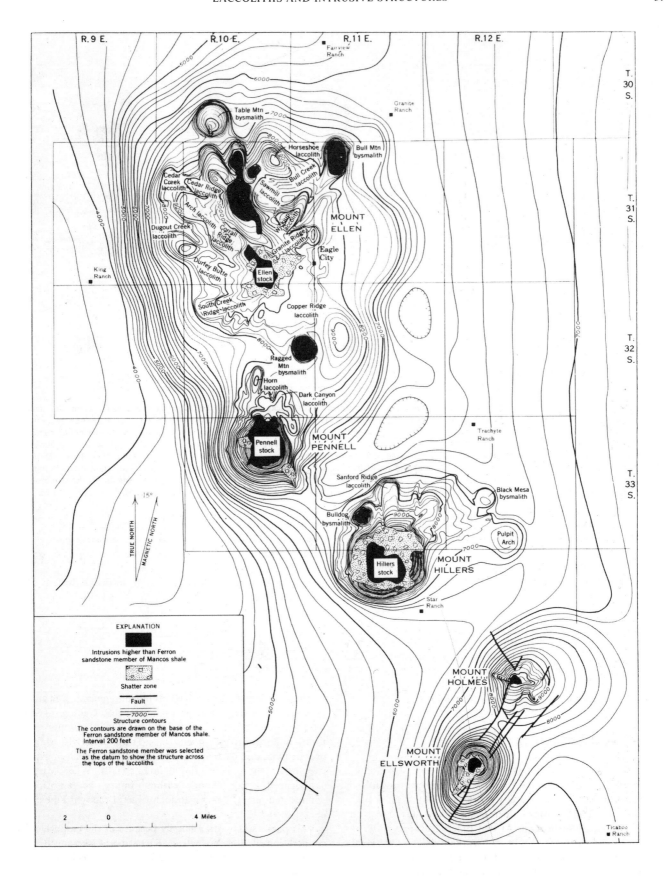

Figure 6. Structure contour map of the Henry Mountains (from U.S. Geological Survey Professional Paper 228, Pl. 5).

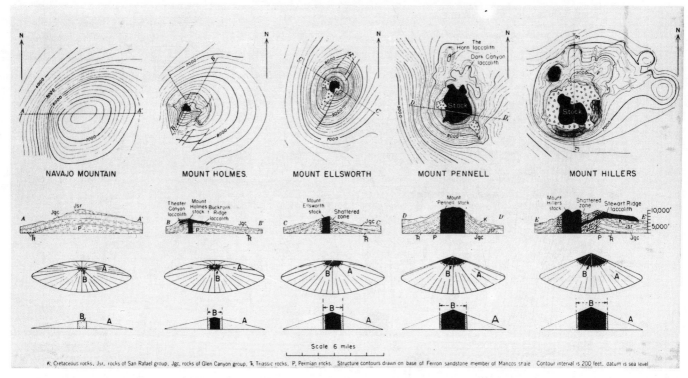

Figure 7. Diagram of the Henry Mountains illustrating the space occupied by the stocks and the theoretical space available for them. The structural geology at Navajo Mountain (after Baker, 1936) and at the four southern Henry Mountains is shown by structure contour maps and geologic cross sections. Below are oblique and cross-section diagrams to illustrate the space available for cross-cutting intrusions that conically deform circular areas. The area of the basal circle of each cone is equal to the area A on the lateral slopes. The area designated B is the amount by which the lateral slope area exceeds the circular base area. In the Colorado Plateau, Navajo Mountain shows that the conditions were such that at least 2,000 ft of uplift in a circular area 6 mi in diameter could be accommodated by stretching the domed strata without creating space for intrusions. Greater uplift in equivalent strata in the Henry Mountains resulted in parting of the strata and creation of space to accommodate the stocks (black). On the cones representing the Henry Mountains, the stippled areas are equal to area B on the cone representing Navajo Mountain (from U.S. Geological Survey Professional Paper 228, Fig. 65).

I discovered those outcrops exposed in small gullies in colluvium otherwise concealing the base of the butte, Gilbert forever became my hero, for it proved he was human.

One potentially important contribution of the resurvey is the recognition that hydrothermal alteration is best developed in and around the stocks rather than around their satellites, the laccoliths and bysmaliths. In the Great Basin, stocks have injected satellitic intrusions, less regular of course than those in the Henry Mountains, but concordant with flat faults or other structural discontinuities. The stocks in such areas can be distinguished from their satellites by their metamorphic halos, and this approach, borrowed from the simpler Colorado Plateau, could lead to mineral discoveries.

SOME FINAL THOUGHTS

The distance from Salt Lake City to the Henry Mountains is some 250 mi, today about a 5-h drive. For Gilbert, traveling by horseback in 1876 at 25 mi a day, it was a 10-day trip!

Another difference is that Gilbert seems to have stayed with one horse most or all of his trip. At that time, feed apparently was plentiful, and a horse could be worked daily. The Henry Mountains region was uninhabited, and there were no livestock. In the next 60 yr, however, the range was badly overgrazed and was devastated when drought struck in the dust-bowl days of the 1930s. Each geologist then needed two horses and alternated riding them.

Even so, supplemental feed was necessary. Now the range is improving, thanks to increasing management since passage of the Taylor Grazing Act. Given proper care, the countryside could be restored to the conditions Gilbert knew.

But the scientific principles about intrusive structures and processes of erosion that Gilbert on horseback brought out of that countryside need no such restoration.

ACKNOWLEDGMENTS

Again I acknowledge the backing, encouragement, and guidance of Hugh D. Miser, Branch Chief of the U.S. Geological Survey, who provided me, then in my late twenties, the opportunity to follow and learn from Gilbert's broad trail. Miser visited the field project twice and participated in that fateful inspection trip when two other bosses were injured in horsefalls. When Miser also experienced a fall, though without injury, he left early and commented later, "Being a Branch Chief, I thought I'd better get out of there."

This paper was written largely because of the enthusiasm and cooperation of Ellis Yochelson. Richard S. Fiske and B. Carter Hearn, Jr., reviewed the manuscript and gave helpful technical suggestions.

My wife, always a partner in my field work and writing, has insisted that I not acknowledge her invaluable assistance.

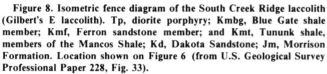

Figure 8. Isometric fence diagram of the South Creek Ridge laccolith (Gilbert's E laccolith). Tp, diorite porphyry; Kmbg, Blue Gate shale member; Kmf, Ferron sandstone member; and Kmt, Tununk shale, members of the Mancos Shale; Kd, Dakota Sandstone; Jm, Morrison Formation. Location shown on Figure 6 (from U.S. Geological Survey Professional Paper 228, Fig. 33).

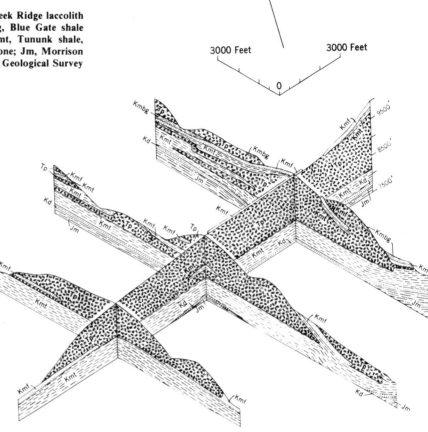

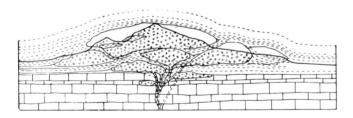

Figure 9. Holmes's concept of the form of instrusions at Ute (El Late) Mountain in southwestern Colorado (from Holmes, 1877). Holmes visualized the laccoliths as having spread laterally from a central source, an interpretation more in accord with later surveys than was Gilbert's (compare Figs. 4, 6, 8).

REFERENCES CITED

Baker, A. A., 1936, Geology of Monument Valley—Navajo Mountain region, San Juan County, Utah: U.S. Geological Survey Bulletin 865, 106 p.

Dana, J. D., 1880, Gilbert's report on the geology of the Henry Mountains: American Journal Science, 3rd ser., v. 19, p. 17–25.

Davis, W. M., 1924, Gilbert's theory of laccoliths [abs]: Washington Academy of Sciences Journal, 14, no. 15, p. 375.

———1925, Laccoliths and sills [abs]: Washington Academy of Sciences Journal, v. 15, no. 18, p. 414–415.

———1926, Biographical memoir, Grove Karl Gilbert: National Academy of Sciences Biographical Memoirs, v. 21, 5th memoir, 303 p.

Emmons, S. F., 1903, The Little Cottonwood granite body of the Wasatch Mountains: American Journal of Science, 4th ser., v. 16, p. 139–147.

Gilbert, G. K., 1877a, Geological investigations in the Henry Mountains, Utah [abs.]: American Naturalist, v. 2, p. 447.

———1877b, Report on the geology of the Henry Mountains: U.S. Geographical and Geological Survey of the Rocky Mountain Region (Powell), 160 p.

———1896, Laccolites in southeastern Colorado: Journal Geology, v. 4, p. 816–825.

Green, A. H., 1879, The geology of the Henry Mountains: Nature, v. 21, p. 177–179.

Gussow, Wm. C., 1962, Energy source of intrusive masses: Transactions of the Royal Society of Canada, 3rd ser., v. LVI, sec. III, p. 1–19.

Hague, Arnold, and Emmons, S. F., 1877, Descriptive geology: U.S. Geological Exploration of the Fortieth Parallel (King), v. 2, 890 p.

Holmes, W. H., 1877, Report on the San Juan district, Colorado: U.S. Geological and Geographical Survey of the Territories (Hayden), Annual Report 9, p. 237–276.

Hunt, Charles B., assisted by Paul Averitt and Ralph L. Miller, 1953, Geology and geography of the Henry Mountains region, Utah: U.S. Geological Survey Professional Paper 228, 234 p.

——1959, About writing reports, some tips from G. K. Gilbert: Journal Geological Education, v. 7, no. 1, p. 1–3.

——1977, Around the Henry Mountains with Charlie Hanks: Utah Geology, v. 4, no. 2, p. 95–104.

Iddings, J. P., 1898, Bysmaliths: Journal of Geology, v. 6, p. 704–710.

Johnson, A. M., and Pollard, D. D., 1973, Mechanics of growth of some laccolithic intrusions in the Henry Mountains, Utah, I: Tectonophysics, v. 18, p. 261–309.

Murchison, R. I., 1839, The Silurian system: London, J. Murray, 576 p.

Neumayr, M., 1887, Erdgeschicte, Bd. 1, Allegemeine Geologie: Leipzig, Bibliographischen Institute, 634 p.

Reyer, Eduard, 1888, Theoretische Geologie: Stuttgart, Schweizerbort (E. Koch), 867 p.

von Buch, M., 1936, Description Physique des Iles Canaries: Paris, 342 p.

MANUSCRIPT RECEIVED BY THE SOCIETY AUGUST 31, 1979
MANUSCRIPT ACCEPTED MAY 20, 1980

Geological Society of America
Special Paper 183
1980

G. K. Gilbert's studies of faults, scarps, and earthquakes

ROBERT E. WALLACE

U.S. Geological Survey, MS 77, 345 Middlefield Road, Menlo Park, California 94025

ABSTRACT

While exploring with the Wheeler Surveys in 1871 and 1872, Grove Karl Gilbert recognized and demonstrated that faulting, rather than folding, dominated mountain building in the basin ranges. Widely accepted now, this concept was challenged by Spurr and others. Gilbert recognized further that relief had been produced incrementally along these range-bounding faults, evidenced by "piedmont scarps," which he first noted and named. Piedmont scarps, in turn, he recognized as evidence of earthquakes, and in 1883 he warned the citizens of Utah that the absence of such scarps along one segment of the front of the Wasatch Range strongly suggested that a large earthquake might eventually occur there. That warning was the first paper on earthquakes authored by a member of the fledgling U.S. Geological Survey.

Gilbert's studies of the San Francisco earthquake of 1906 stand out to present-day investigators as his principal contribution to the knowledge of earthquakes. His photographs, diagrams, and descriptions of the behavior of the San Andreas fault during that earthquake are data that have been used repeatedly during the 1960s and 1970s. But almost unknown to investigators during those decades was Gilbert's paper "Earthquake Forecasting," published in 1909. It was the only paper listed in the *Bibliography of North American Geology* between 1785 and 1922 about earthquake forecasting or predictions. The issues and concepts in Gilbert's paper—earthquake prediction, earthquake engineering, land use, risk evaluation, and insurance—anticipated many elements of the Earthquake Hazard Reduction Act of 1977.

It was not the quality or originality of a particular work of Gilbert that governed its impact on subsequent studies, but rather the existence or nonexistence of a scientific audience, suitably attuned to the subject at hand and sufficiently knowledgeable to perceive and be influenced by his work.

INTRODUCTION

I trespass on fields to which I have no title, for I am an advocate of the principle of scientific trespass. Trespass is one of the ways of securing cross-fertilization for his own crops, and of carrying cross-fertilization to the paddock he invades. [Gilbert, 1909, p. 122]

With these words, Grove Karl Gilbert introduced the most

perceptive, encompassing, and balanced analysis of earthquake hazards and of the means to reduce those hazards that appeared in the United States literature until a series of Federal policy and planning documents of the 1960s and 1970s culminated in the passage of the Earthquake Hazard Reduction Act of 1977. His analysis, contained under the title "Earthquake Forecasts" (1909), was a presidential address to the American Association of Geographers. Although emphasizing that he was not a seismologist, Gilbert had in fact gained a unique understanding of the geologic processes of mountain building which led him inevitably and logically to formulate principles of earthquake mechanisms. His popular article "A theory of the Earthquakes of the Great Basin, with a Practical Application," published in 1884, was the first paper on earthquakes authored by a member of the fledgling U.S. Geological Survey.

During his early explorations with the Wheeler Survey, Gilbert became acquainted with the peculiar mountains of the Utah and Nevada region. He applied the terms "Basin Range System" and "Basin Ranges" to "all that system of short ridges separated by trough-like valleys which lies west of the Plateau System" (1875, p. 22). Later he referred to basin-range structure in describing the structural characteristics of the basin ranges. Evidence there led him to the unorthodox concept that mountain blocks had been raised along fault boundaries, rather than by folds according to the prevalent hypothesis for the Appalachians. The links between mountain building, faults, and earthquakes had been born.

THE BASIN RANGES

Indeed, I entered the field with the expectation of finding in the ridges of Nevada a like structure, (like the folded structure of the Appalachians) and it was only with the accumulation of difficulties that I reluctantly abandoned the idea. [Gilbert, 1875, p. 41]

Gilbert started his studies of the basin ranges in 1871 as a geologic assistant under First Lieutenant George M. Wheeler who was charged with the task of carrying out explorations and surveys west of the 100th meridian. So, at age 28, Gilbert embarked on field studies that led to papers on topics so varied as fluvial lakes, topographic methods, isostasy, intrusion, faults, geomorphology, and earthquakes.

The basin-range system, or basin ranges, provided a wealth of readily accessible new data, for rock exposures abound, physiographic features are unobscured by vegetation, and evidence of dynamic geology and geologically young volcanism is to be seen at every turn. One can only speculate on how modern geologists might have seized upon and used such an opportunity. The wonder of Gilbert's response was his intuitive ability to separate the important from the trivial, to select from the myriad exciting phenomena those observations most significant to the formulation of new concepts, most of which yet survive. One might consider this ability to be precognition, but more likely it demonstrated Gilbert's natural talent for deductive reasoning, which he applied long before he formally published his thoughts on scientific methods (1886). He seems to have had a special talent for perceiving just how far his data could be extrapolated with high probability of being verified by future data.

Gilbert's early reports (1872 and 1875) presented his major and innovative conclusions about the basin-range system: that they were fault-bounded blocks of the Earth's crust; that the forces were uniform in character over large areas; that the forces were nearly vertical and deep seated. Of the contrast to the Appalachians, Gilbert said, "In the Appalachians corrugation has been produced commonly by folding, exceptionally by faulting; in the Basin Ranges, commonly by faulting, exceptionally by flexure" (1875, p. 61).

Gilbert's ability to gain insight from the comparison of large regional differences was undoubtedly enhanced by the rapid reconnaissance nature of the Wheeler and Powell Surveys, which in two years criss-crossed vast areas. Comparing the Colorado Plateau and the basin ranges, Gilbert (1875, p. 61) said,

It is impossible to overestimate the advantages of this field for the study of what might be called the embryology of mountain building. In it can be found differentiated the simplest initiatory phenomenon, not obscured, but rather exposed, by denudation, and the process can be followed from step to step, until the complicated results of successive dislocations and erosions baffle analysis.

Without the advantage of viewing both the Colorado Plateau and the basin ranges in quick succession, perhaps Gilbert would have had greater difficulty in proceeding so surely to the fault-block interpretation of the basin ranges.

In addition to the insight gained from "the embryology of mountain building" seen in the plateau, however, much of the evidence most convincing to Gilbert was within the basin-range system itself. In his very first exposure to the province in April 1871 (Davis, 1926), as he rode the train from Rochester, New York, to San Francisco enroute to join the field party in eastern Nevada, Gilbert must have pondered the amazing structures that he was seeing. In his first major report in 1875 (p. 40), Gilbert described range after range as exhibiting beds in cross section with "but a single direction of dip." Among such ranges were the Pahranagat, Shell Creek at White's Peak, Beaver Creek, Oquirrh, and Thomas. The House Range (Fig. 1) he recognized as having complexities; at its north end, strata dip westward and a precipitous east-facing scarp bounds the range on the east, whereas at Dome Canyon farther south, dips are eastward. He found other ranges such as Bare Mountain and the Amargosa Range to be even more complex. He sketched numerous high-angle faults from afar, and in his notes he expressed frustration at not having the opportunity to examine them closer at hand.

By the reconnaissance nature of his traverses, Gilbert could not have recognized the great thrust-fault structures, whose correct interpretations emerged more than 60 years after his first work.

These structures became clear only after decades of detailed geologic mapping and determination of the ages and facies of rocks regionally. Even more decades of research were required to document the areal extent and amounts of displacements and to test alternative interpretations of the structural patterns that thrust-fault concepts attempt to explain. All anomalies and problems are not even yet resolved, and as geologic data become more abundant, problems seem to multiply. Today, concepts of plate tectonics introduce possible new unifying hypotheses.

Gilbert (1875) considered hypotheses for basin-range topography and structure other than that of faulting. He reviewed the case for denudation of anticlinal and synclinal folds: "Pure anticlinals are exceedingly rare, [although] anticlinals and synclinals also occur as subsidiary features within some ranges" (1875, p. 40, 41). "The simple monocline [homocline] may indeed be explained as the side of an anticline, by the harsh assumption that the remaining parts have been removed below the level of the adjacent valley" (1875, p. 41). He selected Worthington Mountain in eastern Nevada as an example so convincingly fault bounded that little argument seemed needed to convince doubters of the fault boundaries. He said, "I can conceive of no erosion that should have left this thin segment as the remnant of an inclined table or of a fold. Its narrowness, its straightness, and its isolation mark it as a mass of strata thrust upward between two faults" (1875, p. 37).

Despite Gilbert's certainty of the correctness of his interpretation, having compared various hypotheses to his own satisfaction, Josiah Spurr in America and many geologists in Europe did not agree. Spurr (1901, p. 265) took exception even to Gilbert's basic observations: "Ranges consisting essentially of a single monoclinal ridge are exceedingly rare. . .mountain fronts studied are, in general, not marked by great faults." Spurr even derisively stated: "We find them [Gilbert's arguments] almost wholly physiographic." The internal complexity of many ranges dominated Spurr's view. The Funeral Range and Grapevine Mountains of the southwestern Great Basin are clearly more complex than some of the homoclinal ranges that Gilbert first studied. Spurr (1901) stated that "he is glad to call on the opinion of Mr. Clarence King," whom he claimed reported that "monoclinal ridges are in most cases parts of anticlinal and synclinal folds." Perplexed, Gilbert (1928, p. 4) countered, "I fail to discover on that page [referred to by Spurr] any allusion to erosion. On the contrary, King describes certain portions of the Gosiute and Peoquop Ranges as monoclinal and interprets them as the western halves of anticlinals, the entire eastern halves 'having been dislocated downward out of sight'." For some reason Gilbert did not quote the very supportive statement of King that "the frequency of these monoclinal detached blocks gives abundant warrant for the assertions of Powell and Gilbert" (King, 1878, p. 735).

Davis (1903) propounded his own physiographic interpretation of the history of the basin ranges. But while championing Gilbert's use of geomorphic evidence, Davis, in a memorial to Gilbert (1926), criticized Gilbert's statements as incomplete. He complimented Gilbert for including in his theory of the basin ranges "a new and essentially physiographic principle of his own discovery. . . .It led him to take fuller and more reasonable account of surface features than was the habit of the time. Gilbert's contributions to this advance was invaluable; they started a ferment in men's minds." But Davis added that "certain steps of the advance were not clearly apprehended, much less explicitly formulated, by Gilbert in his first studies in the West" (p. 54) and that "Gilbert's theory of the Basin Ranges in its original form must therefore be

Figure 1. West front of House Range, south of Antelope Pass (at left), Millard County, Utah. Photograph by G. K. Gilbert, 1901, upon revisiting the area he reported on in 1875.

regarded as seriously incomplete, so incomplete, indeed, that one may feel surprise at the importance it attained" (p. 65).

The early publication date of Gilbert's concepts, the obscurity of some references, and the dominance in Davis's mind of his own models undoubtedly affected Davis's evaluation in 1926. Gilbert's major, and final, paper on the basin-range system was published posthumously in 1928, two years after Davis's statements in his memorial to Gilbert. Davis's 303-page biographical memorial is testimony that he greatly admired Gilbert. Gilbert also elicited high personal respect even from his adversary Spurr, who in a letter to Gilbert (collection of letters to Gilbert on the occasion of his 75th birthday, on file with U.S. Geological Survey libraries, Denver and Menlo Park) said, "The fact that our opinions have differed on certain matters of physiography and structure does not in the least detract from my admiration of the wonderful work which you have done in building up the science of geology, not only that of American geology, but that of the world."

Gilbert (1928) was pleased to note that G. D. Louderback's description of the faulted and tilted basalt-capped ridges of the Humboldt Lake Range (now the West Humboldt Range) and the Star Peak Range (now the Humboldt Range) reinforced his own interpretation of the basin-range structure: "This relatively local study serves to verify several results of [my earlier] studies and it adds materially to the mental picture of the Basin Range structure" (Gilbert, 1928, p. 7). The basalt caps (now termed "louderbacks") corroborated the evidence of the physiographic form that the mountains were tilted fault blocks.

Gilbert's theories were not without further attack, the most extreme undoubtedly being that of Keyes (1908, p. 33) who postulated that "the origin of the Basin Ranges and of the desert ranges generally is therefore regarded as due in the main to

extensive and rigorous differential deflation on a region that had been previously flexed and profoundly faulted and then planed off." Gilbert noted in 1928 (p. 9) that "up to a certain point the development of opinion was harmonious. . .but the harmony was broken by Spurr and has not since been restored," a disharmony that had at least stimulated Davis and Gilbert to go to the field again for further observations. Gilbert said dryly of Keyes's hypothesis (1928, p. 9): "It remains to be seen whether an equal promotion will result from the equally radical contentions of Keyes."

EARTHQUAKES, FAULTS, AND FORECASTING

In its relation to man an earthquake is a cause. In its relation to the earth, it is chiefly an incidental effect of an incidental effect. [Gilbert, 1912, p. 9].

Gilbert's studies of the devastating 1906 San Francisco earthquake are generally considered today as his principal contribution to the knowledge of earthquakes. His photographs, diagrams, and descriptions of the behavior of the San Andreas fault during that earthquake provide a source of data that has been used over and over again. To a surprising extent, Gilbert's other contributions to the study of earthquakes have gone either unrecognized or underappreciated. In his memorial to Gilbert, W. C. Mendenhall (1920) listed 25 topics upon which Gilbert had written, but failed to refer to "earthquakes." Gilbert is not remembered as the first Survey scientist to author a paper on earthquakes (1884) nor as the only author listed in the *Bibliography of North American Geology* between 1785 and 1922 who wrote a paper on earthquake forecasting.

The currents of thought that influenced Gilbert's concepts are almost impossible to fathom, but his theory of earthquakes of the Great Basin (1884) seems to have been almost entirely his own, from his own observation. Few, if any, of Gilbert's colleagues were considering earthquake problems, except during 1906 and 1907, which may explain why Gilbert's work stood alone, unnoticed, and little affected the evolution of the generally used model of earthquake generation. And yet, Gilbert's theories a century later are almost as complete as are theories diversely created by many scientists over the decades.

At the time of Gilbert's studies, the generally held impression was that surface faulting accompanying earthquakes—so well documented in the Cutch earthquake of 1819 in India (Oldham, 1928)—was a result rather than a cause of earthquakes. In the middle decades of the 1800s, for example, Mallet (1862) believed that strong earthquakes in Italy were fundamentally volcanic in origin. Not until Koto (1893) wrote on the cause of the great Mino- Owari earthquake in central Japan, in which surface faulting was extensive, did seismologists accept the concept that earthquakes were generated by fault displacement. The concept was not fully matured until Reid (1910, 1911) added the elastic-rebound theory, a readily comprehensible physical mechanism by which faulting could generate earthquake waves. But in 1883 Gilbert anticipated these later developments almost completely and in 1909 had further detailed his concepts.

Gilbert's arguments were based on what to him was clear evidence from his own field observations. First, he noted that faults bound the ranges of the Great Basin. When the Earth's crust "breaks, the fracture does not run along the medial axis of the mountains, but along one margin. On one side of the fracture the crust is lifted and tilted; on the other side it either sinks or remains undisturbed." Second, Gilbert's observation of small, fresh scarplets (later to be named "piedmont scarps") at the base of fault-generated range fronts (Fig. 2) indicated to him that

A mountain is not thrown up all at once by a great convulsive effort, but rises little by little. When an earthquake occurs, a part of the foot-slope goes up with the mountain, and another part goes down (relatively) with the valley. It is thus divided, and a little cliff marks the line of division. This little cliff is, in geologic parlance, a "fault scarp", and the earth fracture which has permitted the mountain to be uplifted is a "fault." [Gilbert, 1884, p. 49, 51]

Gilbert was about to start his second year of exploration with the Wheeler Survey when the great earthquake of March 28, 1872, occurred in Owens Valley, California, along the eastern margin of the Sierra Nevada. The fault scarps that developed during that earthquake undoubtedly influenced Gilbert's thinking. By September 1883, his hypothesis was fully matured, and he was compelled to apply it in a predictive, practical sense by warning the residents of Salt Lake City, Utah, of the earthquake hazard there.

In a popular article in the *Salt Lake Tribune*, Gilbert carried the readers through the sequence of logic that (1) mountains of the Great Basin are uplifted along faults, (2) mountains "rise little by little," (3) alternate cohesion and sliding characterizes the motion, and (4) "the instant of yielding is so swift and so abruptly terminated as to constitute a shock" (1884, p. 50). This line of reasoning that relates mountain building to earthquakes is so modern that, in 1980, it is difficult to understand why, once stated, the concept would not have been generally accepted and become a firm part of the working base of geologists and seismologists.

In the nineteenth century, however, the sciences of geology and seismology were unprepared to acknowledge the verity of several major facts that formed the foundation for Gilbert's hypothesis. Few geologists had seen the fault-bounded ranges of the Great Basin. Geologists of the Eastern United States and in Europe conceived tectonism as inferred from the Appalachians and the Alps. Seismologists were preoccupied with the shaking phenomena, not with field observations of the effects of earthquakes. Mallet in Italy was one of the first practitioners of observational field seismology, and it is understandable how awesome volcanic events led him to conclude that volcanism caused earthquakes.

Gilbert recognized that small scarplets were not continuous along part of the Wasatch front.

There is one place where they are conspicuous by their absence, and that place is close to this city [Salt Lake City]. From Warm Springs to Emigration Canyon, faults scarps have not been found, and the rational explanation of their absence is that a very long time has elapsed since their last renewal. In this period the earth strain has been slowly increasing, and some day it will overcome the friction, lift the mountains a few feet, and re-enact on a more fearful scale the catastrophy of Owens Valley. [1884, p. 52]

Here were clear statements of the concept now referred to as "seismic gaps" and of strain accumulation between seismic events. Gilbert's forecast was based in part on the hard fact of a gap in scarplets and in part on a deductive model of a plausible physical mechanism. The forecast is yet unfulfilled.

In discussing earthquake forecasting, Gilbert made a significant distinction between "recognition of premonitory signs" and "recognition of the earliest phases of the event itself" under the general category of what he terms "prelude" (1909, p. 132). Today the distinction between long-term precursors and short-term precursors may represent these two conditions. Gilbert recognized a problem in predicting time of an earthquake that is still unresolved. He says, "I see little practical value in any quality of time precision attainable along lines of achievement now seen to be open" (1909, p. 133). Even today, suitable approaches are vague.

Gilbert went far beyond the areas of geology in which he was well grounded, even beyond the discipline of seismology, in presenting a balanced approach to the reduction of earthquake hazards. He considered and evaluated earthquake risk and the value of insurance; for example, "the minuteness of the earthquake risk may be further indicated by saying that it is one tenth of the risk of death by measles" (1909, p. 136). Use of the risk level of exposure to diseases as a base to which other risks can be compared has come to be a standard approach (Starr, 1969). Gilbert analyzed the risk of earthquakes and compared the feasibility of earthquake insurance to fire insurance, stating that "premiums would be adjusted to local conditions: higher for houses on soft ground than for those on bedrock, relatively high for houses near known earthquake foci, and very low for houses classed as earthquake proof" (1909, p. 137). Gilbert decried the policy of concealment of earthquake hazards, because as he said, "It does not conceal. It is unprofitable, because it interferes with measures of protection against a danger which is real and important" (1909, p. 136).

In summary, Gilbert stated,

It is the duty of investigators—seismologists, geologists, and scientific engineers—to develop the theory of local danger spots, to discover the foci of recurrent shocks, to develop the theory of earthquake-proof construction. It is the duty of engineers and architects so to adjust construction to the character of the ground that safety shall be secured. It should be the policy of communities in the earthquake district to recognize the danger and make provisions against it. [1909, p. 134]

Figure 2. Mountains "rise little by little" and a "little cliff in geologic parlance a 'fault scarp' is formed." Alternate "cohesion and sliding" characterize this motion, "and some day it [strain] will overcome the friction, lift the mountains a few feet, and re-enact on a more fearful scale the catastrophy of Owens Valley." (Top) Fault scarps along the west flank of the Wasatch Range, 1 km north of Little Cottonwood Canyon, Utah. Photographed by G. K. Gilbert, 1901. (Bottom) Same view in 1979 by the author.

PIEDMONT SCARPS AND LAKE BONNEVILLE

The great investigator is primarily and preeminently the man who is rich in hypotheses. In the plentitude of his wealth he can spare the weaklings without regret. The man who can produce but one, cherishes and champions that one as his own, and is blind to its faults. [Gilbert, 1886, p. 287]

Gilbert clearly was gratified and proud of the way in which the puzzles of the Bonneville basin succumbed to his multiple-hypothesis and "analogic" approach. He analyzed his study in his paper "The Inculcation of Scientific Method by Example, with an Illustration Drawn from the Quaternary Geology of Utah" (1886). The Bonneville study required a careful analysis of fault scarps, not only to distinguish them from shore-line features, but also to derive

Figure 3. Piedmont scarps at west base of Wasatch Range, Utah, about 2 km north of Ogden Canyon. "The more recent (scarps), as a rule, stand at the angle of repose for earth, are sharply defined at the top, the less recent are less steep, join the normal piedmont slope by a curve" (G. K. Gilbert, 1928, p. 34).

an accurate assessment of the amount of uplift owed to faulting and to "crust undulation" (isostatic uplift).

As the basin was traversed in the conduct of the general investigation of the old lake, a search was made for the records of recent faults and at every opportunity the height of the shore was accurately measured. It was found that the displacements recorded by the shores have been much larger than the displacements demonstrated by faults. For these reasons faulting was provisionally regarded as a disturbing factor merely. [1886, p. 293]

Gilbert's monograph on Lake Bonneville contains some of the classic descriptions of young faults and a reiteration of his recognition first stated in 1883 that "it is now generally understood that earthquakes are due to paroxysmal yielding of the earth's crust" (1890, p. 360). Strangely, many seismologists and geologists attribute the first recognition that faults cause earthquakes not to Gilbert but to Koto's (1893) description of the 1891 Mino-Owari earthquake. The study of the Bonneville basin began in 1877, when Gilbert, then a member of the Powell Survey, was assigned to study the classifications of lands and the possibilities of irrigation in northern Utah. In a letter sent from Tooele Valley, he noted: "The gauging of streams, the study of beaches, and the study of recent faults go well together and make the greater part of my field work" (*in* Davis, 1926, p. 111).

In 1879 when the U.S. Geological Survey was formed, the original organization included a Division of the Great Basin at Salt Lake City, headed by Gilbert. The Division was abolished in 1883 because of insufficient appropriations to support the Survey operations, which had been extended over the entire United States. But during his four years with the Division and the preceding three years with the Powell Survey, Gilbert amassed the data on which were based not only many of his concepts of faults and fault-scarps but also his classic monograph on Lake Bonneville (1890).

Gilbert's most complete documentation of faulting in the basin ranges did not appear until 1928, ten years after his death, in which he set forth the structural data as "proof" that a fault created the face of many ranges—to the end he was responding to Spurr's challenge.

He elaborated on the "small scarps or bluffs which are evidently results of dislocation. When these were first observed in the Great Basin they were called fault scarps" (1928, p. 33). He felt the need of a "distinctive designation" and chose "piedmont scarp" (1928, p. 34).

He follows with a statement so brief and simple that I had overlooked it until recently and had to rediscover the fact independently. This personal experience exemplifies the wealth of important concepts in Gilbert's writings, which because of their brevity or lack of prominence, have remained cryptic. Thus, "In many places it is evident that the members of a group [of piedmont scarps] were not all made at the same time. The more recent, as a rule, stand at the angle of repose for earth, are sharply defined at the top, the less recent are less steep, join the normal piedmont slope above by a curve" (1928, p. 34) (Fig. 3). The slope of the scarps and sharpness of crests have proved to be very useful criteria of their age and lead to a basis for evaluating paleoseismicity. In rereading the voluminous, concept-packed writings of Gilbert, I have come to wonder how many other nuggets await rediscovery.

ON THE 1906 EARTHQUAKE

It is the natural and legitimate ambition of a properly constituted geologist to see a glacier, witness an eruption, and feel an earthquake. When, therefore, I was awakened in Berkeley on the eighteenth of April last by a tumult of motions and noises, it was with unalloyed pleasure that I became aware that a rigorous earthquake was in progress. [Gilbert, 1907, p. 215]

Three days after the earthquake of April 18, 1906, a State Earthquake Investigation Commission was appointed with Professor Andrew C. Lawson as chairman. G. K. Gilbert became part of a committee to investigate the surface changes, but even before these formalities, the inveterate note taker had recorded in his field notebook:

Wednesday - April 18, 1906
    Berkeley to Oakland and return
    An earthquake shock at 5:11 followed by others at intervals.
    I note 6:11 and 7:47, also 6:53 p.m. Motion in my room N.-S.- in other rooms E.-W. Water spilled from pitcher and chamber. Duration estimated at 1 minute. No apparent damage in laboratory.

Thursday - April 19, 1906
    Berkeley. Unsuccessful attempts to go to San Francisco.
    Mendenhall comes. A faint shock at 10:15 p.m. Booming of dynamite in the city at intervals. The smoke column persists.

Friday - April 20, 1906.
    Berkeley to San Francisco with D. E. Smith and Archsward, Hittel, Healy. Return p.m.
    The made ground about lower Market Street shows much settling, the maximum as ref. to st. car track being 3-4'. The buildings on this ground suffered more from earthquake than those on firm ground.

It was not until April 26 that Gilbert went from Berkeley to Oakland, San Francisco, Ross, and Bolinas, recording in his field notes: "I discover no fallen chimney on Alcatraz." On Marin Peninsula he found that the form of Paradise Valley "suggests that it was determined by a post-plain crack. One of the principal narrow strike valleys I did not reach. It *may* be the principal focus of the quake." Clearly, Gilbert was expecting fault displacement to be the cause of the earthquake, and in his summary for the day he reported: "Some of the cracks were clearly secondary; others may have been primary."

Gilbert continued to follow the San Andreas fault across Point Reyes Peninsula and recorded in his notes of May 10 a diagram of the Skinner ranch and stated that "a path thru the garden, once opposite the front door, is 15' thrown." His diagram shows the now-familiar right-slip displacement along the San Andreas fault. On many days in late April and early May he noted two or three "faint" to "strong" tremors, reporting the time of each as determined from "watch" or "univ. clock."

Gilbert's records of the position and characteristics of surface faulting produced during the 1906 earthquake remain among the best records of such features ever recorded. Even in the 1960s and 70s they were much referred to in understanding the processes of faulting and for evaluating earthquake hazards for land use decisions in California.

Among the many rift features that Gilbert identified were linear valleys, side-hill ridges, elongate ridges, and the depressions that he named "fault sags" (State Earthquake Investigation Commission, 1908, p. 33). He recognized that uplift and tilting accompanied the dominantly strike-slip style of displacement, a feature all too commonly overlooked today because of preoccupation with strike slip.

Gilbert's field notes and writings record his progressive learning, for this was the first major effort in North America to detail characteristics of surface faulting related to an earthquake and the first opportunity anywhere to witness effects of dominantly strike-slip faulting accompanying an earthquake. J. D. Whitney's descriptions of both vertical and strike-slip displacements accompanying the 1872 Owens Valley earthquake, although forming an important base for Gilbert's thinking, did not contain the wealth of detail obtained after the 1906 earthquake. I have difficulty, however, in sensing from the writings of Gilbert's colleagues their predilection for, or opposition to, the concept that earthquakes are fault generated. The concept was little more than a decade old as a major accepted theory, although Gilbert had independently recognized it as a fact more than two decades earlier.

Between Tomales Bay and Bolinas Lagoon, Gilbert classified three phases of the rift topography as the "ridge phase, the trench phase, and the echelon phase" (State Earthquake Investigation Commission, 1908, p. 66) (Fig. 4). He concluded from an analysis of the fractures and ridges that "as the fault-trace is made up almost wholly of these three phases, it follows that in the visible part of the fault its walls did not approach as a result of the faulting but receded a little" (State Earthquake Investigation Commission, 1908, p. 72).

Much of Gilbert's field work was delayed until a year after the earthquake, and in May 1907 Gilbert said, "In the earlier field excursions that fault-trace was here overlookt, the echelon cracks by which it is represented being mistaken for secondary cracks." The delay helped, however, and so "at the present time [spring of 1907] it is easily traced, even from a distance, because the vegetation of the two sides of it has acquired different colors" (1908, p. 66).

Before spring 1907, however, Gilbert must have been preparing the manuscript that was published in August 1907, in which he not only concisely summarized the extent of faulting, but also presented a simple and clear analysis of strike-slip faulting. He also developed the theme, to this day little analyzed and unappreciated as an engineering consideration, of permanent dislocation of ground by wave action. He stated:

Nor do I find any room for doubt either that ridges originated as waves on the surface of the mud while it was rendered quasi liquid by violent agitation, or that they persisted because the mud promptly resumed its normal coherence when the agitation ceased. Whatever their mechanism and history, they illustrate a mode of response of wet, unconsolidated material to powerful earth tremors, and they contribute to an understanding of peculiar destructiveness of the earthquake in such areas. [1907, p. 13]

In addition to studying faulting and ground deformation, Gilbert took copious notes on the intensity of shaking, the overturning of water tanks, shifting of houses off their foundations, and wracking of barns and other structures. He visited Luther Burbank at Sebastopol who gave him an eyewitness account and found that "he could not stand, and settled back against the bed, holding on to the window casing and the bedpost. From the window he saw trees waving, and after the tremor had ceased he seemed to see a continued disturbance in the foot-hills at the east, as though the tremor was retreating in that direction" (State Earthquake Investigation Commission, 1908, p. 205). During the more than 30 years that Burbank resided in Santa Rosa, he had felt about 130 earthquakes, but none was comparable to the 1906 earthquake.

IMPACT ON SUCCESSORS

If a vote were to be taken among American geologists, I believe that G. K. Gilbert would receive an overwhelming endorsement as the leading American geologist of all time, if not the pinnacle of excellence among world geologists. The consistently superb quality of Gilbert's work led T. C. Chamberlain (1918, p. 375) to say in his memorial: "It is doubtful whether the products of any geologist

of our day will escape revision at the hands of future research to a degree equal to the writings of Grove Karl Gilbert."

Gilbert's great variety of studies of uniformly high quality and originality did not have equal impact on subsequent studies. The difference is illustrated by his work on earthquakes as compared to his work on basin-range structure. "Earthquake Forecasts" was never referred to in the deliberations of the 1960s and early 1970s that produced the multidisciplinary approach to a national earthquake hazard reduction program, enacted into law in September 1977. Yet issues and concepts in Gilbert's paper, ranging from earthquake predictions, earthquake engineering, and land use considerations to the evaluation of risk and insurance, anticipated most of the 1977 law. In contrast, Gilbert's ideas about fault-block mountains captured the attention of the scientific community, undoubtedly because of the great interest in mountain building at the time. The controversy with Spurr probably advertised Gilbert's ideas and certainly stimulated Gilbert to document his observations more completely.

Why Gilbert's various works received such different kinds of reception is of interest. A scientific audience must exist and be suitably attuned to the subject at hand in order to perceive and be influenced by a scientific contribution, and the particular science must be dynamic and capable of reflecting changing ideas. "Earthquake Forecasts" fell on deaf ears. By 1909 excitement surrounding the earthquake of 1906 had died away. Until decades later, seismology was poorly funded, the fields of earthquake geology and earthquake engineering were in their infancy, and earthquake economics and sociology were yet to be born. Integration of the whole would not begin for 60 years. No audience or scientific community or program existed to receive and be influenced by Gilbert's work.

Figure 4. San Andreas fault 8 km south of Olema, California. Photograph by G. K. Gilbert, 1906. Three phases of the rift topography are "ridge phase, the trench phase, and the echelon phase" (Gilbert, 1908, p. 66).

Figure 5. G. K. Gilbert photographing a fault surface south of Klamath Falls, Oregon, 1916. See Figure 6 for the photograph obtained by Gilbert. Photograph by J. P. Buwalda from collection of Mrs. J. P. Buwalda. "While the photographers were engrossed in examining the scarp, the brakes of the car in the background released and the car rolled over the road embankment into the bushes," recalled Mrs. Buwalda.

Figure 6. Photograph taken by G. K. Gilbert in 1916 (see Fig. 5) of striae on fault plane 9 3/10 mi south of Klamath Falls, Oregon. "The scratches upon it show a slight variation in direction, and some of them intersect at small angles, but collectively they indicate that the relative motion of the rock masses had no horizontal component within the fault plane" (G. K. Gilbert, 1928, p. 77).

Unused, or not fully used, other concepts of Gilbert undoubtedly lie obscured within the volume of Gilbert's prodigious output. All geologists could profit from repeated study of Gilbert's writings.

POSTSCRIPT ON VERSATILITY

Gilbert's interests, his inquisitiveness, and his versatility were wide-ranging. His geologic photographs of the West must be considered one of his most important legacies. In 17 volumes on file in the library of the U.S. Geological Survey in Denver are more than 3,500 high-quality photographs covering a wide range of geologic subjects. In Figure 5, the photographer Gilbert, then in his seventies, can be seen photographing a fault surface south of Klamath Falls, Oregon (Fig. 6).

On Monday, October 21, 1901, while at Bright Angel Hotel, Grand Canyon, Arizona, Gilbert reported in his notebook: "A lunar rainbow. Moon at 1st ¼ in the south and riding high. Bow in the canyon. Outside the arc uniformily dark. The arc itself pale

gray, without trace of prismatic colors, grading quickly to outer darkness, and very gradually inward."

Gilbert's field sketches demonstrate an artistic sensitivity for line, form, and perspective (Fig. 7). Although his father was a professional portraitist, I find no mention that young Gilbert studied art with him. Many field sketches are rendered on pages subdivided by a grid, as though Gilbert employed the technique of viewing a landscape through a grid as a guide to proportions. The results are most pleasing, yet full of data.

Bird flight captivated Gilbert's imagination, and here again (Gilbert, 1888) he applied his knowledge of physics, his principal formal training, to analyze critical elements of bird flight.

These varied interests and talents epitomize Gilbert's emphasis on facts and observations and the "discovery of their relations. But while the investigator does not succeed in his effort to obtain pure facts, his effort creates a tendency, and that tendency gives scientific observation and its record a distinctive character" (1886, p. 285).

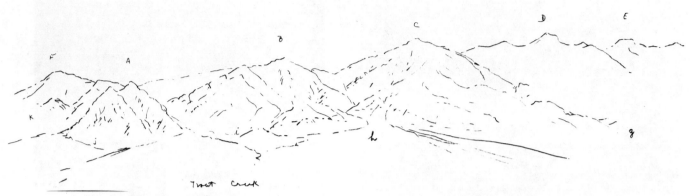

Figure 7. Copied from a field sketch by G. K. Gilbert of the Deep Creek Mountains, September 19, 1901. Trout Creek at left.

REFERENCES CITED

Chamberlain, T. C., 1918, Grove Karl Gilbert [editorial]: Journal of Geology, v. 26, no. 4, p. 375.

Davis, W. M., 1903, The mountain ranges of the Great Basin: Harvard University Museum of Comparative Zoology Bulletin, v. 42, p. 129–177.

——1926, Biographical memoir Grove Karl Gilbert: National Academy of Sciences, v. 21, fifth memoir, p. 1–303.

Gilbert, G. K., 1872, Reports on explorations in Nevada and Arizona [abs.]: Forty-second Congress, second session, Senate Document 65, p. 90–94.

——1875, Report on the geology of portions of Nevada, Utah, California and Arizona examined in the years 1871 and 1872: Report on U.S. Geographical and Geological Surveys West of the 100th Meridian, v. 3, Geology, Pt. 1, p. 17–187.

——1884, A theory of the earthquakes of the Great Basin, with a practical application [from the Salt Lake Tribune of Sept. 30, 1883]: American Journal of Science, 3rd ser., v. 27, p. 49–53.

——1886, The inculcation of scientific method by example: American Journal of Science, 3rd ser., v. 31, p. 284–299.

——1888, The soaring of birds: Science, v. 12, no. 305, p. 267–268.

——1890, Lake Bonneville: U.S. Geological Survey Monograph 1, 340 p.

——1907, The investigation of the California earthquake, in Jordan, D. S., ed., The California earthquake of 1906: San Francisco, A. M. Robertson, p. 215–256/

——1909, Earthquake forecasts: Science N.S., v. 29, no. 734, p. 121–136.

——1912, Preface [to The earthquakes of Yakutat Bay, Alaska, in September 1899, by R. S. Tarr and Lawrence Martin]: U.S. Geological Professional Paper, no. 69, p. 9–10.

——1928, Studies of basin-range structures: U.S. Geological Survey Professional Paper 153, p. 1–92.

Gilbert, G. K., and others 1907, The San Francisco earthquake and fire of April 18, 1906, and their effects on structures and structural materials: U.S. Geological Survey Bulletin, no. 324, p. 1–170.

Keyes, C. R., 1908, Arid monadnocks: Journal of Geography, v. 7, p. 30–33.

King, Clarence, 1878, U.S. Geological exploration of the fortieth parallel, Report v. 1, Systematic geology, 803 p.

Koto, Bundjiro, 1893, On the cause of the great earthquake in central Japan, 1891: Tokyo Imperial University College of Science Journal, v. 5, p. 297–353.

Mallet, Robert, 1862, Great Neapolitan earthquake of 1857. The first principles of observational seismology: London, Chapman and Hall, 2 vols.

Mendenhall, W. C., 1920, Memorial of Grove Karl Gilbert: Geological Society of America Bulletin, v. 31, p. 26–64.

Oldham, R. D., 1928, The Cutch (Kachh) earthquake of 16th June 1819 with a revision of the great earthquake of 12th June 1897: Indian Geological Survey Memoir, v. 46, p. 1–47.

Reid, H. F., 1910, The California earthquake of April 18, 1906, The mechanics of the earthquake: Report of the (California) State Earthquake Investigation Commission: Carnegie Institute of Washington, Pub. no. 87, v. 2, p. 1–192.

——1911, The elastic-rebound theory of earthquakes: California University Department of Geology Bulletin 6, p. 413–444.

Spurr, J. E., 1901, Origin and structure of the basin ranges: Geological Society of America, v. 12, p. 264–266.

Starr, Chauncey, 1969, Social benefit versus technological risk: Science, v. 165, p. 1232–1238.

State Earthquake Investigation Commission, Lawson, A. C., Chairman, 1908, The California earthquake of April 18, 1906: Carnegie Institute of Washington, v. 1, pt. 1 (Gilbert, G. K., author of sections, p. 30–35, p. 66–87, p. 191–198).

MANUSCRIPT RECEIVED BY THE SOCIETY DECEMBER 31, 1979
MANUSCRIPT ACCEPTED MAY 20, 1980

Geological Society of America
Special Paper 183
1980

G. K. Gilbert's Lake Bonneville studies

CHARLES B. HUNT
Visiting Scholar, University of Utah, 2131 Condie Drive, Salt Lake City, Utah 84119

ABSTRACT

Gilbert's reports on Lake Bonneville are, like his Henry Mountains report, classic contributions to geology, but the two studies are very different. The Henry Mountains study involved only 2 months of field work, and the report was completed and published within months after completion of the field work. The Lake Bonneville reports were based on many seasons of field work, first with the Wheeler Survey, then the Powell Survey, and finally the U.S. Geological Survey. Writing and publishing his final report on Lake Bonneville were delayed.

He recognized three main stages of the Pleistocene lake: (1) an early stage represented by what is now known as the Alpine Formation, a major interruption in lake history represented by an unconformity between the early and late lake deposits; (2) another rise of the lake to its highest level, the Bonneville shoreline, overflow of the lake into the Snake River via a gap at Red Rock Pass; (3) drop of water level about 300 ft as the outlet was eroded downward to a bedrock lip, and stillstand of the water at that level (Provo stage) to produce what is known as the Provo Formation. The lake history ended with another 300-ft drop to a poorly developed shoreline (Stansbury stage). The subsequent drop in level of about 300 ft to the level of the Great Salt Lake probably should be considered post-Bonneville history, although Gilbert was vague about the terminal stage.

Gilbert clearly recognized that the lake basin and its islands and peninsular mountains were formed by Tertiary diastrophism. Pre-Bonneville erosion of the mountains produced huge alluvial cones around the mountain bases and partly filled the basins. Lake Bonneville was formed after the alluvial cones and after most of the faulting and volcanism. Faulting, volcanism, and deposition of alluvial cones were renewed during and since the formation of the lake.

Two major contributions to structural geology include recognition of repeated displacements on faults along the Wasatch Front and doming of the lake basin as a result of isostatic rise due to unloading of the crust as the lake desiccated. Both structural contributions have been amply confirmed by modern surveys.

INTRODUCTION

Gilbert was 32 years old when he named Lake Bonneville in 1875. In the same report (Gilbert, 1875, p. 88–89) he named the Bonneville beach, the Provo beach, and applied the name "Bonneville group" to the lake deposits. This was a year before his classic study of the Henry Mountains and 15 years before his monograph on Lake Bonneville was published (Gilbert, 1890).

He named the lake in honor of Captain B. L. E. Bonneville who, in 1833, gave the first authentic account of the Great Salt Lake (see Washington Irving, 1848). In a footnote in his first report, Gilbert stated (1875, p. 89) that Captain Bonneville "saw Great Salt Lake in 1833." Whether Bonneville ever saw the lake is uncertain, but at least he prepared the first map of it, even if only from accounts given to him by members of his exploring partly.

Gilbert's field work and reporting on the Henry Mountains and Lake Bonneville projects are interesting contrasts. His Henry Mountains report (1877) was based on 60 days in the field, and his report was written, transmitted, and published within months after returning to the office. In present-day parlance, it was a "quickie." In contrast, his Lake Bonneville report was based on many seasons of field work, and the project was interrupted, as modern studies are likely to be, by administrative chores and changes in assignment.

Gilbert had reconnoitered the Bonneville basin and first reported on it while he was with the Wheeler Survey (Gilbert, 1875). His more detailed follow-up research began with the Powell Survey and was continued during the early years of the U.S. Geological Survey. Like so many modern reports, publication came long after the field work ended—7 years. But that early record has been broken many times since.

Actually, the first monograph of the U.S. Geological Survey was supposed to be *Precious Metals of the United States* by Clarence King (see announcement at beginning of Monograph II by Dutton, 1882). King's monograph never appeared, and during the next 8 years, 14 other monographs were published before it was decided to fill the vacant number 1 spot with Gilbert's final report on Lake Bonneville.

The writing, editing, and assembling of Gilbert's Lake Bonneville monograph do not compare favorably with his Henry Mountains "quickie." The Lake Bonneville monograph is lengthy, partly because it covers a large area, but partly because geographical and topical reporting are mixed. As a result, important information on structural geology is scattered in four places (Gilbert, 1890, p. 211–213, 263–265, 331–332, and 340–392); fossils are listed in three places (p. 209–211, 297–305, and 393–402), and geochemistry is discussed in four places (p. 167–169, 200–208, 225–228, and 251–258). The illustrations are distributed almost at random through the text, many far removed from the pages they illustrate. This lack of coordination makes reading difficult.

Like many (most?) present-day authors, Gilbert experienced difficulty "keeping up with the literature." His preliminary reports on Lake Bonneville had to have details modified when he discovered that certain problems being discussed by him had been considered previously by others. Most geologist today feel frustrated trying to give adequate reference to previous work. Gilbert experienced the same frustration a hundred years ago.

Gilbert paid little attention to the economic potentialities of either the Henry Mountains or the Lake Bonneville basin. He dismissed the former with the pithy statement, "No one but a geologist will ever profitably seek out the Henry Mountains" (Gilbert, 1877, p. 14). His Lake Bonneville report casually mentions only two deposits of potentially economic importance: salt around the Great Salt Lake and gypsum in dunes in the Sevier Basin (Gilbert, 1890, p. 253, 323).

Water supplies and irrigable lands in the Great Salt Lake basin received much of his attention. These data are not included in his Lake Bonneville reports but are given in Powell's report on the arid lands (Powell, 1879, chaps. IV, VII).

While studying Lake Bonneville, Gilbert saw and described the bottom effects of more lake and shore processes than can be seen or sampled from a modern, expensively equipped research vessel. Utah is far inland and remote from the coasts and Great Lakes, but the Lake Bonneville basin with its Great Salt Lake and fresh-water bays provides a unique laboratory for studying limnology, including sedimentation and erosion, lake-bottom conditions, salinization, ecology, wildlife refuges, and land utilization in deserts. Why not a Gilbert Institute of Desert Basins at one of Utah's universities?

GILBERT IN THE FIELD

The several thousand miles traveled by Gilbert and his assistants while studying the Lake Bonneville basin were mostly by horseback and wagon. Some travel, however, was "by the cars" (railroad cars), and clearly travel by railroad in those days was informal. His notebook for October 25, 1877 [or 78? Some one has changed Gilbert's 1877 to 78] recorded, "Train behind please bring over my mail from post office."

His notebooks for September 26 to October 4, 1879, logged his first travels for the U.S. Geological Survey, an 8-day trip by railroad from Washington via Jackson, Michigan, to Salt Lake City. This was 9 years after the first Pullman Hotel Express had crossed the country. Gilbert had to borrow $1,000 to begin his 1879 field season.

Other field trips "by car" included visits to the various settlements along the railroad north of the Great Salt Lake. In January 1880, after traveling by horseback, with camp moved by wagon, from Salt Lake City to the southwestern corner of the Great Salt Lake Desert and Snake Valley in westernmost Utah, his party returned to Deseret (near Delta, Utah), and from there Gilbert returned to Salt Lake City by railroad. Today, only a hobo can make that trip. That Gilbert did not travel as a hobo is indicated by his expense account which included $1.65 for collars and tie.

His travels by horseback and wagon had advantages, however, for one does see more detail and traces more contacts laterally by saddle than can be done through an automobile windshield. Such plodding travel not only gave Gilbert a closer view and a *feel* for the

ground, but also provided time for him to reflect about features being observed. Ideas could be checked as he rode along the embankments and camped at springs below the shorelines. There was no temptation to delude himself into believing that his ideas could be thought out later. He could weigh and record his evidence and hypotheses as field work progressed. But this mode of travel had its inconveniences too. For example, one day on that trip back to Deseret, he recorded, "Thompson and I miss course to Coyote Spring and do not reach camp until 4 A.M."

Illustrations in Gilbert's monograph are sketches rather than photographs (Fig. 1), and many or most were drawn by talented assistants, although Gilbert continued to sketch landscapes (Fig. 2) as he did in the Henry Mountains. Drawing scenes in the field to emphasize geologic relationships is almost a lost art, replaced by photography; yet there is a place for both, because drawings made in the field require closer observation and more thought than does pointing a camera. Earth science departments would do well to stress landscape drawing in the field because it helps in understanding the geology. Utilizing the camera's advantages need not deprive one of the advantages of older methods.

Gilbert's notes also revealed that mules were of greater value than horses. While negotiating for his needs for traveling the benches from Logan to Salt Lake City (October 30, 1879), he could obtain a wagon with cover, a driver, and harnesses. This would cost $125, including two mules and four horses, but only $115 if all were horses.

Both in his Henry Mountains and Lake Bonneville studies, Gilbert had time only to sample the areas being studied, which is further tribute to his ability to observe and to analyze.

SOME OF GILBERT'S CONTRIBUTIONS

"When the work of the geologist is finished and his final comprehensive report written, the longest and most important chapter will be upon the latest and shortest of the geologic periods" (Gilbert, 1890, p. 1).

This prediction is increasingly being realized. However, studies of surficial deposits, erroneously called "Pleistocene" or "geomorphic studies," went through a period of doldrums during the first third of this century when emphasis was diverted to studying landforms (geomorphology) instead of deposits. Geologists still confuse geomorphology with physiography despite the difference in scope and meaning of those terms as originally used and as defined in unabridged dictionaries. Gilbert's study of Lake Bonneville was physiographic and not merely geomorphic, for it included water budgets, geochemistry, structural geology, paleontology, and, as already noted, land classification for irrigation.

One of Gilbert's major contributions to geology, developed from his Lake Bonneville studies, was not included in his monograph but was published elsewhere (Gilbert, 1886). It is an analysis of the role of the hypothesis in an inexact science like geology. He wrote (p. 286),

A phenomenon having been observed. . .the investigator invents an hypothesis in explanation. He then devises and applies a test. . . . If it does not stand the test he discards it, and invents a new one. . . . He continues [testing] until he finds an hypothesis that remains unscathed.

But, although Gilbert contributed greatly to Quaternary geology in both his Henry Mountains and Lake Bonneville reports, he equated "Quaternary" with "Pleistocene" (1890, p. 22, 396). He

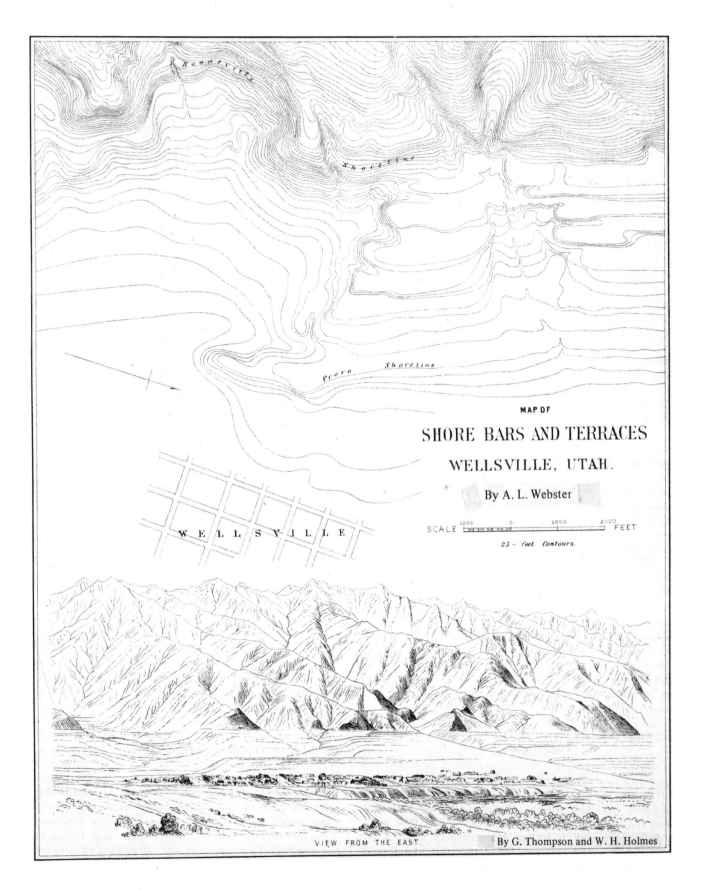

Figure 1. Map and diagram illustrating shore bars, terraces, and mountain front near Wellsville, Utah, showing contrast between the subaerial topography above the Bonneville shoreline and the littoral topography below it (from Gilbert, 1890, Pl. XII).

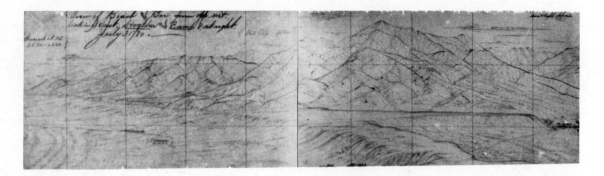

Figure 2. Sketch of Stockton bar and beach ridges, looking east (from Gilbert's notebook). See Figure 2A of Pyne (this volume) for a clearer sketch of this area.

failed to identify the Recent, now known as Holocene, although it had been defined clearly by Lyell in 1863 (p. 5): "In the Recent we may comprehend those deposits in which not only all the shells but all the fossil mammalia are of living species. . . ." Gilbert did passingly refer to this fact of paleontology at the very end of his Lake Bonneville report (p. 401–402) in writing that the mammalian fauna of the Great Basin experienced great change at the close of the Pleistocene. Yet even today, many geologists try to treat the Quaternary differently from any other system by avoiding paleontology as the basis for its stratigraphy. The otherwise excellent summary of the Holocene in the 1974 *Encyclopaedia Britannica* (v. 8, p. 998–1007) does not even mention differences between Pleistocene and Holocene mammalian faunas.

TOPOGRAPHIC FEATURES OF LAKE SHORES

When Gilbert moved from Salt Lake City to Washington, D.C., in the spring of 1881, his field work was interrupted, but he obtained access to more complete library facilities. He found that several European scholars had anticipated him in writing about the topography of shorelines. Some are listed in the bibliography by Johnson (1919, p. 83–86). European interest in shorelines was inspired mostly by interest in navigation. Gilbert (1885, 1890, p. 25) arrived at his ideas independently, and his viewpoint based on "fossil" shores was unique. In discussing littoral erosion, transportation, and deposition, he was one of the first, if not the first, to recognize the importance of the unusual storm, referred to today as the "once in a hundred years storm." Also, he distinguished internal sedimentary structures of the deposits. In deltas, for example, he distinguished what today are called "topset, foreset, and bottomset beds."

LAKE BONNEVILLE SHORE FEATURES

Gilbert emphasized the contrast between the landforms produced by stream erosion above the Bonneville shoreline and the horizontality of the landscape characterized by beach lines below the Bonneville (Fig. 1). Above the Bonneville shoreline are alluvial cones and above them "all is of solid rock" (Gilbert, 1890, p. 91). Actually, bare rock is far less extensive even in the desert mountains than most geologists, including Gilbert, have appreciated; more than 80% is mantled by colluvium or other surficial deposits.

Gilbert correctly noted (1890, p. 93) that "the subaerial work

antecedent to the lake. . .has greatly exceeded. . .the lacustrine work; and the last has. . .exceeded the subaerial work subsequent to the lake. . . ." Expressed differently, the effects of the Holocene have been minor compared to those of the Wisconsinan, and the effects of the Wisconsinan have been minor compared to those of the pre-Wisconsinan Pleistocene.

Because it separates two very different landscapes, the Bonneville shoreline is ". . .conspicuous by reason of its position. . . .It insists on recognition" (Gilbert, 1890, p. 94). The emphasis is expressed attractively. The Bonneville shoreline is about 1,000 ft higher than the Great Salt Lake. Lake Bonneville's area was almost 20,000 mi^2 and its coast more than 2,500 mi long (Fig. 3). It was roughly two-thirds the size of the largest of the Great Lakes, as Gilbert noted (1890, p. 106).

Whereas the Bonneville shoreline is conspicuous because of its position, the next lower one, the Provo (Fig. 3),

is rendered conspicuous chiefly by the magnitude of its phenomena. Its embankments are the most massive, and its wave-cut terraces are the broadest. Moreover, the Provo lake was in every way inferior to the Bonneville. . .for the generation of powerful waves. It was narrower and shallower and obstructed by larger islands. To have constructed shores equal to those of the Bonneville, it must needs have existed a longer time; and still longer to have built the great structures. [Gilbert, 1890, p. 127]

All the cities along the Wasatch Front are built on deltas of the Provo stage of Lake Bonneville. He named the stage for the city of Provo (Gilbert, 1875, p. 90). Gilbert accounted for the paucity of Bonneville-level deltas by noting that the Bonneville level extended into valleys draining from the mountains; sediments deposited there were eroded from the valleys when the lake fell to the Provo level. Those reworked sediments and fresh additions built the extensive Provo-level deltas on the gentle slopes below the mouths of the canyons.

Gilbert noted (1890, p. 132) that at many places one could recognize substages of the Provo, "but all efforts to correlate them and deduce a consistent history have failed." This is still true, but during recent years *formal* stratigraphic names applied by the U.S. Geological Survey to the substages have proliferated. The classic Provo comprising those deposits forming the big deltas and bars in places is not even recognized. The newly named units have not been traced around the basin; they are local. Until systematic mapping shows they are not attributable to local storms or to faulting, I urge retaining Gilber's terminology and introducing refinements by *informal* names for the substages. It should also be noted that the "new stratigraphy" includes Holocene deposits with the Bonneville. Are the deposits of last summer's flash floods Bonneville?

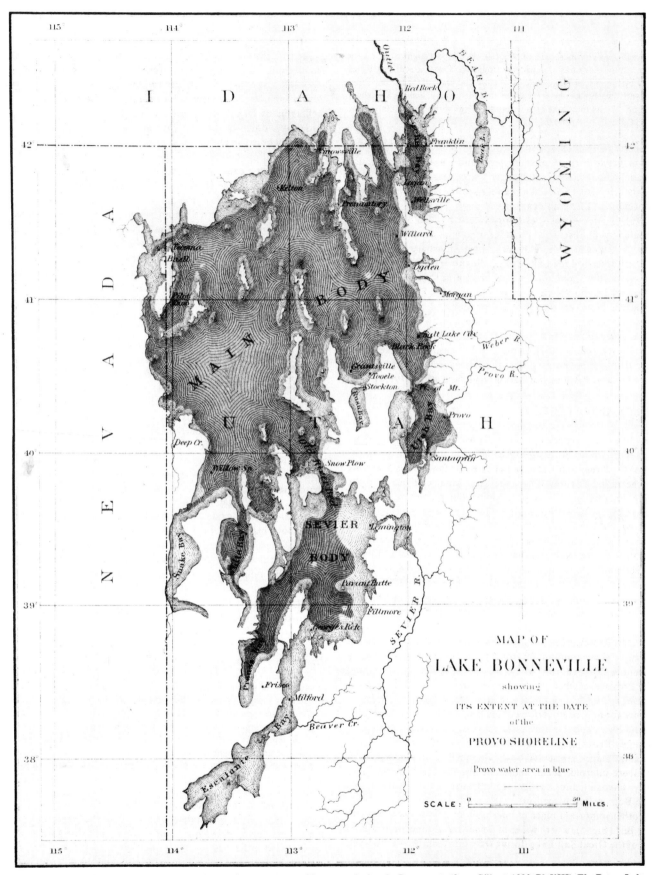

Figure 3. Map of Lake Bonneville showing its maximum extent and its extent during the Provo stage (from Gilbert, 1890, Pl. XIII). The Provo Lake covered about 13,000 mi³, "11,500 belonging to the main body and 1,500 to the Sevier body" (Gilbert, 1890, p. 134).

[Although]...a considerable number of shores can be recognized below the Provo...only...one has been widely recognized. That one is distinguished merely by the greater magnitude of its cliffs and embankments, but is not sufficiently accented to be everywhere identified. It is called the Stansbury shoreline. . . . Its total lake area was in the neighborhood of 7,000 square miles. [Gilbert, 1890, p. 134-135]

Gilbert gave little attention to shore features in the 300-ft interval between the Stansbury level and the Great Salt Lake. Archaeological evidence at Danger Cave, near Wendover, indicates that the Great Salt Lake has not risen 100 ft since the cave was first occupied 11,000 radiocarbon years ago (Jennings, 1957, p. 90, Fig. 70). That the post-Stansbury interval is Holocene is substantiated by the absence of Pleistocene mammalian fossils in caves below the Stansbury shoreline or in post-Stansbury deposits.

Gilbert described the Lake Bonneville shorelines with respect to their geomorphic position, and referred to the shorelines between the Bonneville and Provo levels as "Intermediate Shore-lines." He recognized, however, that the "Intermediate" deposits are older than either the Provo or Bonneville deposits which overlap them. "In the order of time the Intermediate comes first and the Provo last" (Gilbert, 1890, p. 157). In order to emphasize the stratigraphy rather than the geomorphology, the "Intermediate" deposits are referred to as the Alpine Formation (Hunt and others, 1953, p. 17).

OUTLET AND OVERFLOW

By Gilbert's interpretation, after deposition of the Intermediate shorelines (Alpine Formation), the lake level fell, perhaps to the level of the Great Salt Lake, as indicated by fan gravels overlying what he called "Yellow Clay," the deep water equivalent to the Alpine Formation. These post-Alpine fan gravels in turn are overlain by the Provo Formation. The lake, by his interpretation, next rose to the level of the Bonneville shoreline, overflowed to the Snake River at Red Rock Pass (Fig. 4), cut rapidly downward into alluvial fill at the pass, and then maintained its level at a ridge of bedrock which determined the Provo level of the lake.

Gilbert's interpretation of the history at Red Rock Pass has been questioned; it seems difficult to accept the premise that the lake rose so close to the level of the pass without overflowing when the Alpine Formation was deposited. On the other hand, it is difficult to account for the abrupt change from the fine-grained deposits of the Alpine at steep mountain fronts to coarse gravels representing the Bonneville stage. And why is there no evidence for a stillstand at the Provo level when the assumed Alpine-Bonneville lake fell to the level of the bedrock ridge at Red Rock Pass? Yet there is evidence for only one overflow of the lake north of Red Rock Pass (Malde, 1968), and it is difficult to visualize no overflow during the prolonged Provo stage. The critical evidence still is missing.

That the stage represented by the Bonneville shoreline was brief, as Gilbert inferred, seems indicated also by the fact that later surveys have failed to identify lake-bottom sediments attributable to the Bonneville Formation.

Another important place of interbasin overflow was at the Old River Bed (Fig. 5), where water of Sevier Bay drained to the main body of the Great Salt Lake during recession from the Provo level.

OSCILLATIONS OF LAKE BONNEVILLE

Outcrops that Gilbert took as type section for Lake Bonneville bottom deposits are along the Old River Bed "near...where it is crossed by the Overland Stage road" (Gilbert 1890, p. 189-190). The locality has historical as well as geologic interest and is accessible even by modern automobiles. The top of Gilbert's section includes: (1) recent eolian sand; (2) lacustrine sand with shell remains like those in units 3 and 4; about 10 ft thick and gradational downward with 3; (3) white marl, a fine limey clay that weathers white; gypsiferous base crowded with shells; thickness 10 ft—erosional unconformity; and (4) yellow clay, laminated, olive gray on fresh exposure but weathers pale yellow; sand lenses; gypsiferous; shells like those in the overlying beds; thickness more than 90 ft. This base is not exposed.

It has been noted above that no one has yet found lake-bottom beds that could be equated with the Bonneville shoreline. The base of Gilbert's unit 3 is "crowded with shells"; similarly the base of the lake-bottom members of the Provo Formation in Utah Valley are crowded with shells (Hunt and others, 1953, p. 23-25). The occurrences seem to record a unique ecological condition. Might that be the deep-water stage represented by the Bonneville shoreline? If so, one might look for a similar zone rich in shells in the Alpine Formation.

Laterally toward the mountains in the general area of the Old River Bed, Gilbert found fan gravels at the erosional unconformity between the clay and the marl, further confirming this as the major break in the lake history. He also found, as others have done, alluvial gravels intertongued with other high-level parts of the lake-bed section showing there had been oscillations of lake level, but he did not let such minor occurrences and fluctuations of lake level obscure the fundamental bipartite history of the lake. Such obfuscation did not come until nearly 100 years later when, as already noted, later studies cast unnecessary darkness on what had been light.

Gilbert found that the marl contained more carbonate and sulfate and less chloride than the yellow clay. He also found that the yellow clay contained more ferric oxide. The yellow clay (Alpine Formation) probably was derived in large part from erosion of thick, deeply weathered and deep red pre-Wisconsinan loess in the mountains, which source could account both for the high proportion of clay in the Alpine Formation and for its color (Hunt and others, 1953, p. 42).

Shortly before Gilbert began his Lake Bonneville studies, Mark Twain crossed the Old River Bed on an Overland Stage Coach. He did not share Gilbert's enthusiasms, and his account of the desert to the west is very different. Mark Twain wrote (1871):

Imagine a vast, waveless ocean stricken dead and turned to ashes; imagine this solemn waste tufted with ash-dusted sage-bushes; imagine the lifeless silence and solitude that belong to such a place; imagine a coach, creeping like a bug that went by steam; imagine this aching monotony of toiling and plowing kept up hour after hour, and the shore still as far away as ever, apparently; imagine team, driver, coach and passengers so deeply coated with ashes that they are all one colorless color; imagine ash-drifts roosting above mustaches and eyebrows. . . .

Gilbert listed the molluscs and other fossils found in Lake Bonneville deposits and stated (1890, p. 210) that there was only one extinct form, *Amnicola bonnevillensis*, but that species was not included in his list.

How thick are the Lake Bonneville beds and the older fill in the basin? Gilbert did not attempt to answer the first part of the question other than to note that the lake beds were at least 110 ft thick at the Old River Valley. If the log of a well drilled at Saltair has been correlated correctly, the Lake Bonneville beds in that area

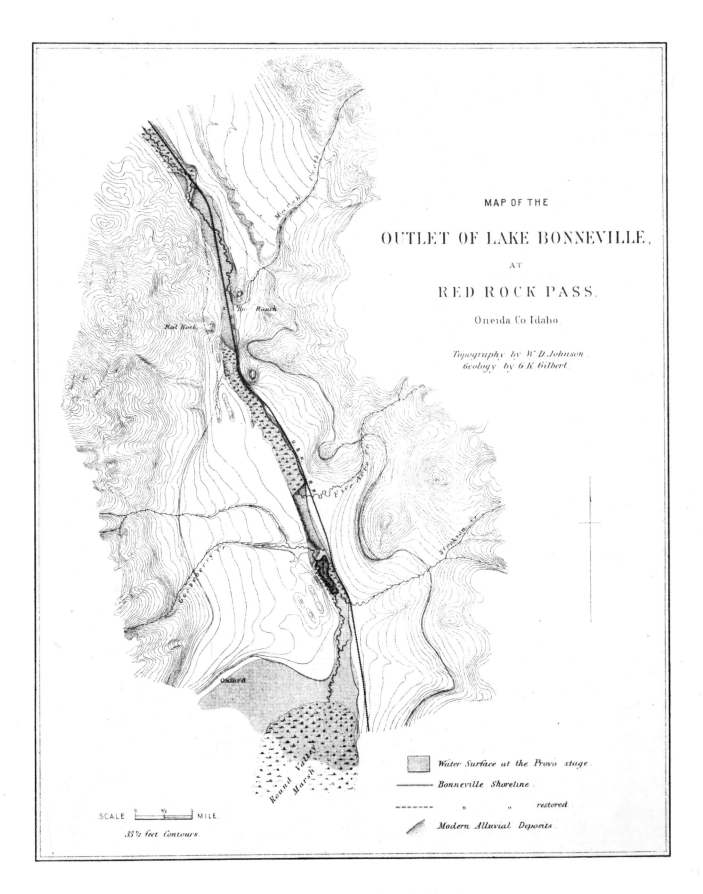

Figure 4. Map of the outlet of Lake Bonneville at Red Rock Pass (from Gilbert, 1890, Pl. XXVIII).

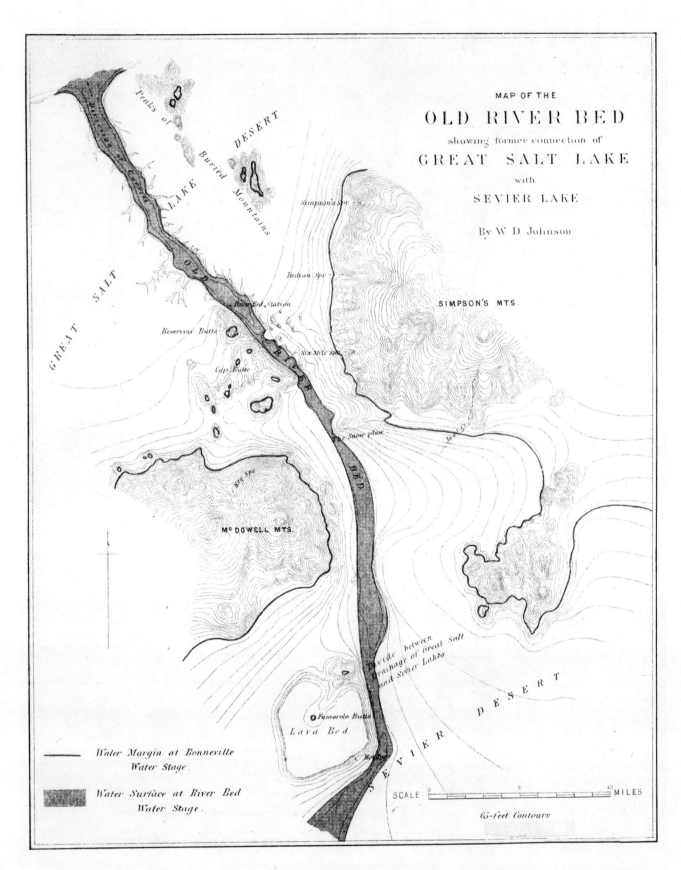

Figure 5. Map of the Old River Bed showing former connection of Great Salt Lake with Sevier Lake (from Gilbert, 1890, Pl. XXXI).

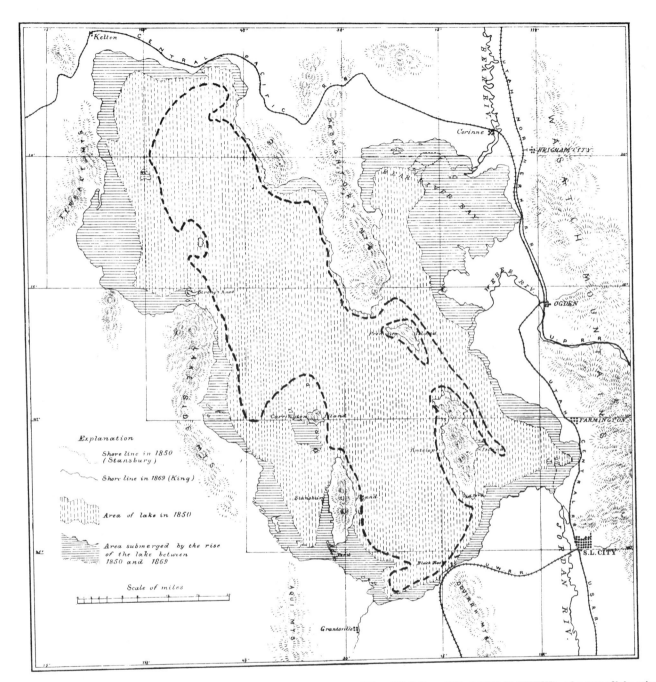

Figure 6. Map of Great Salt Lake, Utah, showing its increase in area between 1850 and 1869 (from Gilbert, 1890, Pl. XXXIII) and extent of lake at its lowest historic level (1963). Present level is about the same as in 1850.

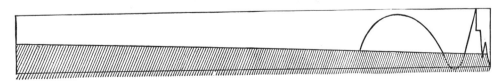

Figure 7. Fluctuations of water in Lake Bonneville (from Gilbert 1890, Fig. 34). First rise (rounded curve) is that of the Intermediate (Alpine) stage. The second peaked at the Bonneville shoreline, receded to the Provo, and then to the Stansbury. Recent studies have added minor irregularities, which Gilbert had noted, but he also correctly pointed out that none could be traced around the basin, and none has been. "The shaded area represents ignorance" (Gilbert, 1890, p. 262).

Figure 8. Pavant Butte, a volcanic cone in the Sevier Basin, has shorelines eroded on the old parts of the cone, but Gilbert found young eruptions there that overlapped lake beds. View from the southwest (from Gilbert, 1890, Pl. XL).

are about 400 ft thick (Eardley and Gvosdetsky, 1960). Gilbert (1890, p. 215) estimated total fill of more than 2,000 ft because mountains in the Great Salt Lake Desert for example, (Newfoundland Mountains) are submerged in lake Bonneville beds and their alluvial bases are buried. "The mountains. . .conform to the Great Basin type in the characters of their summits, but are almost devoid of alluvial cones" (Gilbert, 1890, p. 215). He noted that much of the fill is pre-Pleistocene.

HISTORY OF THE BASIN

Gilbert recognized that the Lake Bonneville basin was due to diastrophism, antedating the Pleistocene. The alluvial cones, he noted (1890, p. 220) are older than the lake, and their development was interrupted when the lake formed. Gilbert also considered the post-lake history, concentrating on the historic record from 1845 to 1883.

Figure 6, from his monograph, has practical applications today, for it shows the tremendous changes in area of the Great Salt Lake

caused by slight changes in depth. It covered 1,750 mi[2] in 1850 and 2,170 mi[2] in 1869. Subsequent changes recorded are 2,400 mi[2] in 1873, only 950 mi[2] in 1963, and 1,600 mi[2] in 1970.

He attributed the rise (to 1869) to increased runoff because of livestock grazing in the mountains and increased agriculture with irrigation that involved developing springs and runoff from what had been marshes. Today we have two added complications: increased runoff from the roofs and pavements of the urban corridor along the Wasatch Front and diversion of water from the Colorado River system to maintain lawns and water-loving trees in what was desert along the urban corridor.

Gilbert's reconstruction of historic changes in level of the Great Salt Lake is recorded in his field notes from accounts of residents, some of whom used rocks or stakes as gauges. Others used more equestrian measures, recalling dates when submerged places could be forded; one ford was described as "saddle deep on a horse 10 hands high."

Changes in level of the Great Salt Lake also affect its salinity. As Gilbert wrote (1890, p. 251–252), "The lake is so shallow that its volume is greatly affected by small changes in level, and since the

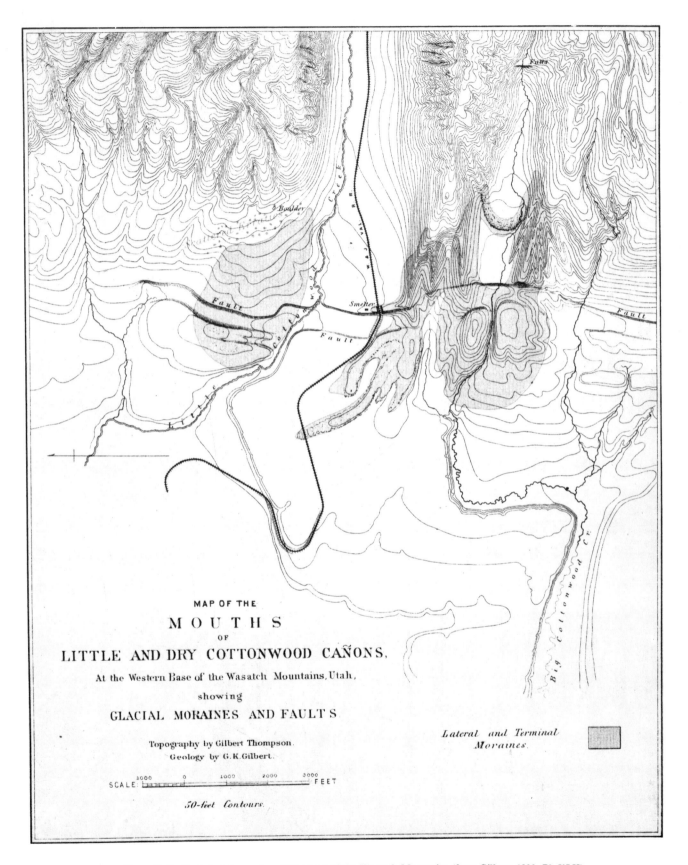

Figure 9. Faulted glacial moraines at the foot of the Wasatch Mountains (from Gilbert, 1890, Pl. XLII).

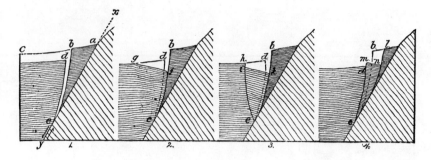

Figure 10. Diagrams illustrating alternative mechanisms for developing grouped fault scarps like those in the unconsolidated deposits along the Wasatch Front (from Gilbert, 1890, Fig. 48).

Figure 11 (facing pages). Gilbert's structure contour map showing doming of Lake Bonneville shorelines compared with the structure contour map based on modern surveys. (Left) Gilbert's map, 50-ft contours, datum Great Salt Lake. (Right) Map based on modern surveys, 40-ft contours, datum sea level (after Crittenden, 1963; 20-ft contours in the original omitted for easier comparison with Gilbert's map).

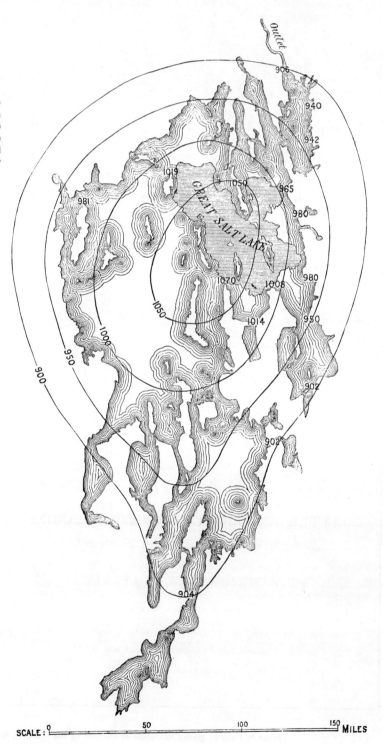

total amount of salts undergoes no appreciable change, the strength of the solution is affected." With the lake at a low stage in 1850 "Stansbury collected a sample. . .containing 22.4 per cent solid matter. From a sample gathered in 1873 when the lake was high. . .the salinity was decreased by 39 per cent. . . . The volume of water in the lake was at the same time increased 73 per cent."

Comparing recharge of various salts, Gilbert wrote (1890, p. 255–257), "The streams carry enough calcium to charge the lake to the observed extent in 18 years, but 34,000 years are necessary to

similarly charge it with chlorine." Balancing recharge and loss of various salts is complicated by the fact that the salts are precipitated in shallow water edges of the lake and in wet muds during recessional stages.

Figure 7 summarizes Gilbert's interpretation of the rise and fall of Lake Bonneville. He found stratigraphic and geomorphic evidence for two major stages of flooding and showed further that these were preceded by an even longer period when the alluvial cones were formed. He related the lake and glacial deposits in

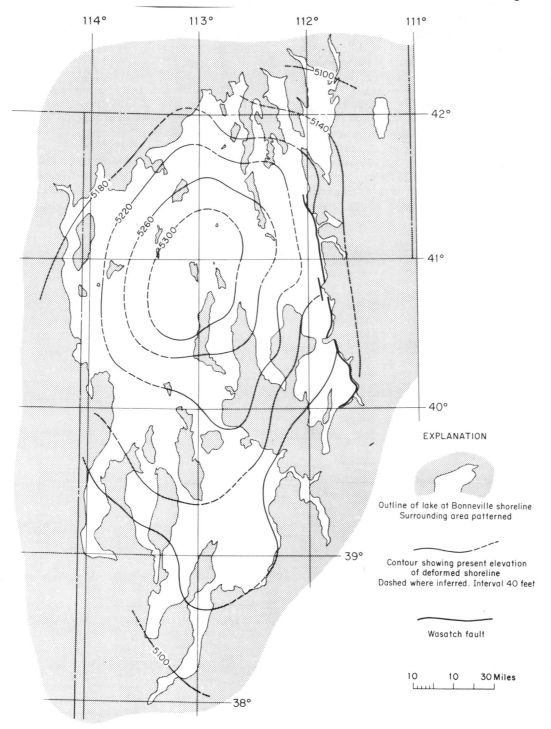

EXPLANATION

Outline of lake at Bonneville shoreline
Surrounding area patterned

Contour showing present elevation
of deformed shoreline
Dashed where inferred. Interval 40 feet

Wasatch fault

10   10   30 Miles

southern Salt Lake valley and, finding that the molluscan fossils from Lake Bonneville deposits average smaller than the same species still living in Utah Lake, he attributed this to dwarfing due to cold.

VOLCANISM

Most volcanic eruptions in the Bonneville basin occurred before the basin was flooded by the lake (Gilbert, 1890, chap. VII). Gilbert included the lava and hot spring at Fumerole Butte (Fig. 5), now considered Pleistocene. In the southeast part of the Sevier Basin, he found basaltic lavas interbedded with the lake sediments, shorelines eroded on older parts of some volcanic cones and younger lavas from these cones that overlapped lake beds.

Cones like Pavant Butte (Fig. 8) that have scoriae heaped highest on the east or northeast sides indicate that southwesterly winds were prevailing when the cones erupted.

Other basaltic lavas in the Bonneville Basin arc more weathered than the ones interbedded with lake beds, and Gilbert cited this as evidence that they are pre–Lake Bonneville and mostly Tertiary. However, he equated Lake Bonneville with Pleistocene rather than with the last (Wisconsinan) glaciation of the Pleistocene, and some of his pre–Lake Bonneville lavas would now be classed as Pleistocene.

CONTINUED DIASTROPHISM

Gilbert also found (1890, chap. VIII) that most of the faulting and other diastrophic changes were pre–Lake Bonneville but that the faulting continued during the history of the lake and has continued since. He found that the faults displaced glacial moraines as well as the lake beds (Fig. 9) and noted (p. 341) that the young displacements represented renewed movement on older ones.

Along many of the faults, Gilbert found grabens and interpreted them as illustrated in Figure 10. He stated that post-Bonneville faulting amounted to about 40 ft (1890, p. 359). Being familiar with the Owens Valley earthquake of 1872, he cautioned that Salt Lake City and other cities along the Wasatch Front were vulnerable to renewed earthquakes. His warning, expressed a century ago, is still sound. Today the potential for major loss is even greater because of urban sprawl across the still active Wasatch fault zone and onto the steep mountain front.

One of the major contributions of Gilbert's Lake Bonneville study was demonstrating isostatic rise of the lake basin almost certainly, as he hypothesized, as a result of unloading of the lake water from that segment of the crust. He discovered the doming early in the course of making aneroid barometer surveys of the shorelines, and he was one of the first, if not the first, to use structure contours to show displacement.

Figure 11A is Gilbert's map of the domed lake basin. Figure 11B is the same made 83 years later. Gilbert's "horse and wagon" map compares favorably with the later one based on modern leveling, aerial photography, and topographic maps. The total discrepancy is less than 10% (Crittenden, 1963, p. E8). Gilbert cited his hypothesis of isostatic rise as the example of the role of hypothesis for seeking explanations that cannot be proved.

Gilbert attributed the easterly position of the Great Salt Lake to the eastward tilting there. Later work in the Great Salt Lake

Desert, where chlorides are crowded toward the west, suggests westward tilting there (Nolan, 1927, p. 40). Comparing the weight of the volume of lake water and the amount of rebound led Gilbert to estimate crustal rigidity. He apparently used the term "isostasy" in an early draft of his report, because the word appears in the index of his monograph but is not used in the text.

EQUUS BEDS

Gilbert differed with Cope about the age of deposits containing *Equus* and associated vertebrates that now are extinct. Cope (1879) argued that they were Pliocene. Gilbert showed that they were correlative with the Great Basin lacustrine deposits, correlative with the glaciations, and therefore were Pleistocene. He also commented on the extinction of certain vertebrates at the "close of the Pleistocene" (1890, p. 400, 402). Still, as already noted, he did not distinguish Pleistocene from Recent (Holocene), as Lyell had done and as implied in Gilbert's usage in Monograph I.

Some day a radiometric, absolute chronology of the stages of lake Bonneville will become available, but that day is not here yet. A tabulation of *all* published radiocarbon dates gives only a wild scatter of numbers, and we do not know which are wrong or why. Until we know Lake Bonneville stratigraphy in all parts of the basin and have correctly correlated the radiometric dates with that stratigraphy, we will do best by staying with the general picture developed by Gilbert. After all, he saw more sections in more parts of the basin and with a broader viewpoint than any of us who have looked more intensively at parts of the basin. We need to identify those members of the Provo Formation that are attributable to local storms or to local structural movements. The tendency today, in adding refinements, is to look at the trees, whereas Gilbert was looking at the forest.

REFERENCES CITED

Cope, E. D., 1879, The relations of the horizons of extinct Vertebrata of Europe and North America: U.S. Geological and Geographical Surveys of the Territories (Hayden), Bulletin 5, p. 33–54.

Crittenden, M. D., Jr., 1963, New data on the isostatic deformation of Lake Bonneville: U.S. Geological Survey Professional Paper 454-E, p. E1–E31.

Dutton, C. E., 1882, Tertiary history of the Grand Canyon district, with atlas: U.S. Geological Survey Monograph 2, 264 p. and atlas of 21 sheets.

Eardley, A. J., and Gvosdetsky, V., 1960, Analysis of Pleistocene core from Great Salt Lake: Geological Society of America Bulletins v. 71, p. 1323–1344.

Gilbert, G. K., 1875, Report on the geology of portions of Nevada, Utah, California, and Arizona: U.S. Geographic and Geological Explorations and Surveys West of the One Hundredth Meridian (Wheeler Survey), v. III, Geology, p. 17–187.

——1877, Report on the geology of the Henry Mountains: U.S. Geographical and Geological Survey Rocky Mountain Region (Powell Survey), 160 p.

——1885, The topographic features of lake shores: U.S. Geological Survey 5th Annual Report, p. 69–123.

——1886, The inculcation of scientific method by example, with an illustration drawn from the Quaternary geology of Utah: American Journal Science, v. 31 (3rd ser.), p. 284–299.

——1890, Lake Bonneville: U.S. Geological Survey Monograph I, 438 p.

Hunt, Chas. B., Varnes, H. D., and Thomas, H. E., 1953, Lake Bonneville:

Geology of northern Utah Valley, Utah: U.S. Geological Survey Professional Paper 257A, 99 p.

Irving, Washington, 1848, The adventures of Captain Bonneville: Chicago and New York, Belford, Clark and Co., 300 p.

Jennings, Jesse D., 1957, Danger Cave: Memoir of Society of American Archaeology, v. XXIII, no. 2, pt. 2, 328 p.

Johnson, D. W., 1919, Shore processes and shoreline development: Reprinted by Hafner, 1965.

Lyell, Charles, 1863, The geological evidences of the antiquity of man: Philadelphia, George W. Childs, 518 p.

Malde, Harold E., 1968, The catastrophic late Pleistocene Bonneville flood in the Snake River Plain, Idaho: U.S. Geological Survey Professional Paper 596.

Nolan, T. B., 1927, Potash brines in the Great Salt Lake Desert, Utah: U.S. Geological Survey Bulletin 795-B, p. 25 to 44.

Powell, J. W., 1879, Report on the lands of the arid region of the United States, with a more detailed account of the lands of Utah: U.S. Geographic and Geological Survey of the Rocky Mountain Region (Powell Survey), 195 p.

Pyne, Stephen J., 1980, A great engine of research—G. K. Gilbert and the U.S. Geological Survey, *in* Yochelson, Ellis L., ed., Scientific ideas of G. K. Gilbert: Geological Society of America Special Paper (this volume).

Stansbury, H., 1852, Exploration and survey of the valley of the Great Salt Lake of Utah: For the U.S. Senate, Philadelphia, Lippincott, Grambe and Co., 487 p.

Twain, Mark, 1871, Roughing it: American Publishing Co., reprinted Harper and Bros., 1913, p. 127.

Manuscript Received by the Society September 17, 1979
Manuscript Accepted May 20, 1980

Geological Society of America
Special Paper 183
1980

Pioneering work of G. K. Gilbert on gravity and isostasy

DON R. MABEY

U.S. Geological Survey, Room 508, Post Office Building, Salt Lake City, Utah 84101

ABSTRACT

G. K. Gilbert's major works relating to gravity and isostasy consisted of three separate efforts: (1) a study of the deformation of the Lake Bonneville shorelines, (2) participation in the first gravity profile across North America, and (3) a U.S. Geological Survey Professional Paper on the interpretation of gravity anomalies. With the Lake Bonneville data, Gilbert established that the strength of the Earth's crust in the Great Basin could not support the load of water in Lake Bonneville and that the basin subsided and rebounded as the water load was first applied and then removed. As a participant in the first major gravity survey in the United States, he experimented with techniques of interpreting gravity observations and speculated on the significance of these sparse data. As one of his last major scientific undertakings, he wrote an essay calling for a more flexible approach to the interpretation of regional gravity variations and outlining his thoughts on the deep structure of the Earth. Although each of these three efforts made a significant scientific contribution, the conclusions he reached in each are not totally consistent with one another, and he attempted neither to reconcile the inconsistencies nor to develop a unified theory of isostasy. Apparently by hypothesizing in each instance, he hoped to stimulate the geodesists to a broader view in their examination of the data.

Gilbert's extraordinary talents are apparent in his work in gravity and isostasy, but he does not appear to have had a great impact on the geodesists who had dominated the development of these disciplines for many years. However, his conclusions that major deformations of the crust reflect "horizontal movements of the upper rocks (lithosphere) without corresponding movements in the nucleus and thereby imply mobility in an intervening layer (asthenosphere). . ." and that the driving tectonic forces stem from a "primordial heterogeneity of the earth which gives diversity to the flow of heat energy. . . ." (Gilbert, 1914, p. 34, 35) indicate a remarkable insight into large-scale tectonic processes.

INTRODUCTION

G. K. Gilbert's work in the field of gravity and isostasy is not as well known as his other major activities, but he published eight papers that centered upon these subjects in a major way. He identified and described the evidence for isostatic rebound of the floor of the Lake Bonneville basin, Utah; he participated in the first major gravity survey in the United States; and the subject of one of the last scientific efforts he completed was the interpretation of gravity anomalies. An examination of Gilbert's works relating to gravity and isostasy provides an insight as to the breadth of this remarkable scientist's view as well as a commentary on the development of earth science in the United States.

Although the term "isostasy" was coined in 1889 by a colleague of Gilbert's (Dutton, 1889), the concept is much older. In the sixteenth century, Leonardo da Vinci recorded in his notebook an explanation of why mountains continued to rise, and thus he may be regarded as the first isostasist (Delaney, 1940). In the middle of the eighteenth century, Bouguer (1749) determined that the measured gravitational attraction of the Andes was much smaller than that computed, and Boscovich (1750) postulated that this was a reflection of a low density mass at depth compensating the mass of the mountain. The next major advance did not occur until one hundred years later when Pratt (1855) published his model explaining isostatic compensation of the Himalayas based on measurements of the deflection of the vertical. Pratt concluded that mountains were forced up by the process of expansion producing a corresponding attenuation of matter below. A month later the astronomer Airy (1855) proposed a quite different model, one that called for mountain roots, and compared the state of the Earth's crust lying upon the lava to the state of a raft of timber floating upon the water. Pratt's model required a lateral density decrease beneath mountains, whereas Airy's required a thickening of the crust. Several geodesists in the last half of the nineteenth century confirmed the existence of compensating low density masses underlying large elevated areas, but the two basic models envisioned remained those proposed by Pratt and Airy. Development in the 1880s of pendulums capable of accurately measuring the Earth's gravity field added another dimension to the geodetic studies.

LAKE BONNEVILLE REBOUND

G. K. Gilbert's concern with isostasy may have begun on that day in 1872 when he discovered the regional deformation of the Lake Bonneville shorelines and started a search for its cause. Gilbert's classic work on Lake Bonneville is the subject of another paper in this volume; here we are concerned with only one aspect:

his conclusion as to the cause of the variation in the altitude of the Provo and Bonneville shorelines. Gilbert (1890a, p. 373) observed that the maximum elevation of the shorelines was in the central part of the Lake Bonneville basin and concluded that "this coincidence suggests the hypothesis that the disappearance of the Lake and the epeirogenic rise of the center of its basin stand in the relation of cause and effect."

In isostasy Gilbert was to be an investigator in a field largely dominated by theorists. In a paper read before the Society of American Naturalists, of which he was president in 1885, he used his work on deformation of the Lake Bonneville shorelines as an example of a proposed approach to scientific investigations and stated the conclusions that were later published in the Lake Bonneville monograph. His essay "The Inculcation of Scientific Method by Example" was presented to the Society of American Naturalists in December and published the following year in the *American Journal of Science* (Gilbert, 1886). In this remarkable essay, Gilbert wrote of

the prime difference between the investigator and the theorist. The one seeks diligently for the facts which may overturn his tentative theory, the other closes his eyes to these and searches only for those which will sustain it. . . . Evidently, if the investigator is to succeed in the discovery of veritable explanations of phenomena, he must be fertile in the invention of hypotheses and ingenious in the application of tests. [Gilbert, 1886, p. 286]

This philosophy was to be demonstrated in his later studies of isostasy.

In U.S. Geological Survey Monograph I, Gilbert (1890a) first described the deformation of the shorelines and then set out on a search for an explanation. Initially he considered the effect on the geoidal surface of removing the water and concluded that

The engineer's level should not find the Bonneville shoreline higher on the central islands than on peripheral slopes. The theoretic change corresponds in kind with the observed; does it agree in amount? My mathematical resources not being adequate to this question, it was submitted to my colleague R. S. Woodward, who gave it full consideration.

The conclusion was that the maximum change of the geoidal surface would be 2.01 ft, which was much less than the 168 ft he observed.

Next he explored the effect of warming. He assumed that the water temperature at the bottom of Pleistocene Lake Bonneville was the temperature for the maximum density of water, $39°F$, and that for 100,000 yr of post-Bonneville time, the mean annual temperature had been $51°F$. He again turned to Woodward for help in computing the rise of the Earth's surface due to thermal expansion, and the result was 1.28 ft, which was also much too small.

Having eliminated these two hypotheses, he then concentrated on loading and unloading as an explanation.

Gilbert examined a model with a solid crust resting on a liquid substratum. Assigning limiting densities of 2.75 and 3.50 g/cm^3 to the liquid substratum, he computed the limiting heights of arching of the Bonneville shoreline to be 364 and 286 ft, as compared to the measured deformation, which he assumed for this computation to be 129 ft (Crittenden [1963b] later determined that the maximum deformation was actually 210 ft). Gilbert then used an "engineer's formula" to compute that a beam 31.7 mi thick would be required to "stand the strain" indicated by the difference between the measured rebound and the computed limit. He did not take into

consideration in his computations the possibility that the viscosity of the liquid substratum might have prevented the basin from achieving either the maximum subsidence or rebound.

Gilbert recognized that the data on Lake Bonneville provided an excellent opportunity for further study and suggested:

A thorough treatment is on the one hand highly desirable and on the other beset with difficulties. It is desirable because it promises to throw some light on the condition of the interior of the earth; a solid earth would not yield the same deformation as an earth partly liquid; a highly rigid earth would behave differently from one of feebler rigidity. It is difficult because it must deal with magnitudes and pressures far beyond the field of experimentation, and can be accomplished only by the aid of comprehensive mathematical analysis. It requires an analytic theory of the strains set up by a stress applied locally to the surface of the earth and of the resulting deformation, and this theory must be so general as to include divers assumptions as to the variation of elasticity with depth from the surface, and as to the relation of the strains to the limits of elasticity. The evolution of such a theory is beyond my power, but in the belief that it is worthy of the attention of the mathematician and physicist, I will endeavor to state the problem.

He then proceeded to state the problem primarily as a study of rigidity, without considering the information to be obtained on the mobility of the substratum. Gilbert identified the opportunities presented for a comprehensive study of the deformation in his Lake Bonneville monograph and it is surprising that more than 70 yr elapsed before anyone (Crittenden, 1963a, 1963b; Walcott, 1970) took on the problem and published the next major studies of the phenomenon. Walcott assumed a density for the displaced material of 3.3 g/cm^3 rather than Gilbert's 3.5 g/cm^3 and used Crittenden's data that indicate a rebound of 65 m rather than the 40 m Gilbert assumed in his computation. Walcott's computed thickness of the rigid crust was about 20 km; Gilbert's was 50 km.

The organization of the Lake Bonneville monograph is somewhat puzzling. Following the section "Hypothesis of Terrestrial Deformation by Loading and Unloading," which appears to reach a final conclusion, Gilbert presented new data in a section titled "Evidence from the Position of Great Salt Lake." He then followed this with a section "The Strength of the Earth," which appears to be essentially an afterthought. He began:

The writer has been led by the discussion of these phenomena to a conception of the rigidity or strength of the earth, more definite than he had previously entertained. It would not be proper to call this conception a conclusion from the data here presented, or a result to which they rigorously and necessarily lead. It is rather a working hypothesis suggested by the study of Lake Bonneville. [Gilbert, 1890a, p. 387]

In the discussion that followed, Gilbert made clear that his thinking had evolved considerably from the earlier monograph sections on deformation. Now he recognized the problem of viscous flow and the role of time, stating:

The viscous flow will consume time, and when it has ceased, there will remain a system of elastic strains. Beyond the elastic limits, the laws of change for loading the surface of the earth (and similarly for unloading) are quasi-hydrostatic. [Gilbert, 1890a, p. 388]

Gilbert then attempted to use his observations of Lake Bonneville and the Wasatch Range to determine the load that the rigidity of the Earth could support, assuming that the Wasatch Range was supported by rigidity but that Lake Bonneville was not.

He wrote:

The phenomena of faulting at the base of the Wasatch, whether considered by themselves or in connection with the filling of the adjacent valley with water and its subsequent emptying, appear to my mind best accordant with the idea that the Wasatch Range and the parallel ranges lying west of it are not sustained at their existing heights above the adjacent plains and valleys by reason of the inferior specific density of their bases and the underlying portion of the crust, but chiefly and perhaps entirely in virtue of the rigidity or strength of the crust. [Gilbert, 1890a, p. 388]

He was not completely correct. The smaller ranges to the west are chiefly supported by the strength of the crust, but the Wasatch Range itself, although apparently moving independently of the subsidence and rebound of Lake Bonneville, is not. Gravity surveys (Cook and others, 1975) made 60 yr after Gilbert's study revealed a large Bouguer gravity low over the Wasatch Range and high plateaus to the east relative to the Lake Bonneville basin. The Wasatch Range is part of an area of high terrane east of the basin that is not supported chiefly by the strength of the crust but is buoyed up by a mass deficiency at depth. Gilbert had determined that a mass the size of the Lake Bonneville water load, which he estimated to be equivalent to a rock load of 730 mi³, was not supported by the strength of the Earth's crust. However, he had not established that the crust could support a load the size of the Wasatch Range, which by choosing rather arbitrary boundaries he estimated to have volume of only 200 mi³. He also noted:

We can not deny the equal possibility, first, that the strength of the earth varies so widely in different places that a measure discovered in the Bonneville basin serves merely to indicate the order of magnitude of a measure of the average strength. . . .

Essentially the same material as that contained in the section "Strength of the Earth" in the monograph was Gilbert's first communication to the newly organized Geological Society of America in December 1889. Titled "Strength of the Earth's Crust" (Gilbert, 1890b), this paper was published in the first *Bulletin* of the Geological Society of America.

Gilbert was working with the Lake Bonneville data while Dutton was developing the concepts of isostasy, which he defined in his paper "On Some of the Greater Problems of Physical Geology" (Dutton, 1889). This paper was published one year before publication of the Lake Bonneville monograph, and it was followed by a discussion by Gilbert (1892). Gilbert made a passing reference to Dutton in the Lake Bonneville monograph and even included the word "isostasy" in the index, with reference to page 387. However, the term "isostasy" is not used on page 387 nor anywhere else in the text. Apparently the index was prepared last and received the term as a measure of currency, without corresponding modification of the text.

While Gilbert was making his study of Lake Bonneville, others were investigating deformation related to the continental ice sheets. Gilbert (1890a), in discussing the mathematical work done by Woodward, stated, "It happened that the cognate problem of the deformation of the geoid by a continental ice mass was submitted to him at about the same time by Dr. T. C. Chamberlin, and he was thus led to a comprehensive discussion of the general subject to which the special problems belong." In the discussion of Gilbert's paper (1890b), Branner asked Gilbert if he believed that the ice load was responsible for "the northward depression of this country during the glacial epoch." Gilbert replied, "I regard the

hypothesis as most valuable, and one that will stimulate investigation. It is too early yet to accept it or reject it. I may say that it is my own working hypothesis. . . ." (Gilbert, 1890b).

With the publication of U.S. Geological Survey Monograph 1, Gilbert (1890a) had established that the Earth's crust in the Great Basin could not support the 300-m-deep and 200-km-wide lake that had filled the Bonneville basin and that the basin had subsided under the load and then rebounded when the load was removed. He concluded correctly that material underlying a rigid crust had flowed out and back to accommodate the subsidence and rebound. He identified as worthy of the attention of the mathematician and the physicist an area of research that would shed light on physical properties of the interior of the Earth. He further warned that the response of the Earth to loading might vary widely from place to place.

1894 GRAVITY MEASUREMENTS

By 1894 the U.S. Coast and Geodetic Survey had developed their half-second pendulum to the point that they were ready to establish a transcontinental profile of gravity measurements. T. C. Mendenhall, Superintendent of the Coast and Geodetic Survey, approached Gilbert and asked and received Gilbert's help in selecting locations for the gravity stations because Mendenhall "desired to have the points of observation so chosen as to give light not only on geodetic problems but also on problems of peculiar interest to geologists" (Gilbert, 1894). A decision was made to have Gilbert, who would be working during the summer of 1894 in Colorado, make an examination of the geology around some of the gravity stations. In the spring and summer of 1894, G. R. Putnam of the Coast and Geodetic Survey established 26 pendulum stations, most of them near the 39th parallel east of Salt Lake City. Two stations had been previously established in California. The exact location of the stations was communicated to Gilbert in Colorado, and he visited 10 of the stations enroute from Colorado to Washington. Gilbert supplied Putnam with a description of the geology of the stations and estimates and measurements of the density and thickness of the rocks underlying the stations.

The formulas for the reduction of gravity data had been developed by 1894, but the significance of the reduced data was not well understood. Putnam computed free-air anomalies, complete Bouguer anomalies, and Faye anomalies.[1] Putnam gave Gilbert the principal facts for at least the 10 stations that Gilbert had visited and the results of part, but not all, of Putnam's reduction of the gravity data.

Gilbert (1894) wrote a letter on March 19, 1895, through the Director of the U.S. Geological Survey, to General W. W. Duffield, who was at that time Superintendent of the Coast and Geodetic Survey, reporting the results of his work with the gravity data. The Director of the Geological Survey, Charles D. Walcott, in a transmittal letter (Gilbert, 1894) said, "As this work on behalf of the Geological Survey was merely accessory to that of the Coast Survey, it seems proper that its results should be published by the Coast Survey in conjunction with the report of its own officer." Gilbert's letter was published along with Putnum's (1894) report as

[1] A Faye anomaly is the free-air anomaly adjusted for the attraction of a layer with thickness equal to the difference between the elevation of the station and the average elevation surrounding the station; it is similar to an isostatic anomaly, but no model is assumed.

part of a progress report of the Coast and Geodetic Survey's work for the fiscal year ending with June 1894. (Although this report bears an 1894 date, it was not published until 1895; it contains letters written in 1895.) Gilbert's letter is concerned primarily with the effect of applying a correction to the gravity anomalies for the difference between the measured density of the rocks underlying the 10 stations and the average density of the crust assumed in reducing the data.

Three days before the date of Gilbert's letter report to the Coast and Geodetic Survey, he read the paper "Notes on the Gravity Determination Reported by Mr. G. R. Putnam" to the Philosophical Society of Washington. This paper, which was published by the Philosophical Society (Gilbert, 1895a), revealed that Gilbert had done much more with the gravity data than described in the letter to Duffield. A footnote indicated that he had not seen Putnam's manuscript when he wrote the manuscript published by the Philosophical Society, but Gilbert did read Putnam's published report before his own manuscript was printed.

Whereas Putnam computed anomaly values as we use them today (the difference between measured gravity and computed gravity), Gilbert corrected the measured gravity to a standard latitude and sea level. He then computed what he called "residuals" in the letter to Duffield and "departures" in the Philosophical Society paper. These values for each station were the difference between the corrected value at a station and the mean of all the corrected values. Gilbert reasoned that if all the corrections were properly applied, the corrected values for all the stations would be equal. He found that if he applied only the free-air correction, the mean deviation of his departures was 0.009 dyn.[2] Applying the geological correction to the free-air anomalies did not change the the mean deviation of the departures. Applying the Bouguer correction (the effect of the mass of material between the station and sea level) produced a mean deviation of the departures of 0.064 dyn. The geological correction reduced this mean deviation to 0.054 dyn. Gilbert's (1895a, p. 65) conclusion:

So far as the discussion of ten stations may warrant a general conclusion, that conclusion would appear to be that corrections based upon the densities of the accessible rocks may be applied with advantages to observations discussed under the theory of high rigidity, but not to observations discussed under the theory of isostasy. It is well to hold lightly a generalization from so small a number of particulars, but this one is strengthened by certain theoretic considerations. Under the isostatic theory the mass of each unit column of earth matter is approximately the same, differences of mean density being compensated by differences of altitude, and variations from the mean in one part of the column being compensated by opposite variations in other parts. Allowance for the density peculiarities of a part should not therefore be made. Under the theory of high rigidity there is no compensatory adjustment, and correction may properly be made for any local peculiarity which is discovered.

In current terminology, Bouguer anomaly values are usually more useful than are free-air anomalies in studying local mass anomalies.

It was apparent to Gilbert that the important elevation to consider in relating measured gravity to elevation was not the singular elevation of the point of observation but rather the average elevation of the terrane surrounding the station. He thus saw the need of another correction—one that would account for

the effect of a layer of rock equal in thickness to the difference between the station elevation and the average elevation of the Earth's surface around the station. This correction applied to the free-air anomaly provides a value approximately equal to the average free-air anomaly for the area and the free-air anomaly for a station at the average elevation. Gilbert called this "the adjustment to the mean plain." Putnam called it "Faye's reduction," and the anomaly value thus computed has been called a "Faye anomaly." Gilbert averaged elevations within 30 mi of the station, and Putnam averaged elevations within 100 mi. Although this reduction was replaced by isostatic reductions in the early 1900s, the method has advantages in some applications over the more elaborate isostatic reductions (Mabey, 1966).

Gilbert computed a mean of the gravity values after reduction to mean plain for all of the stations of what he called the "interior plain" (his original 10 stations plus a station at Ithaca, New York) and assumed that this area was in approximate isostatic equilibrium and thus the mean value could be considered as a standard. He then took the remainder of Putnam's stations and determined the difference between the measured value after reduction to the mean plain and the mean for the standard stations. This value he used to compute the thickness of a sheet of rock that would produce the difference. He used the unit "rock-feet" for this value.

Although Gilbert recognized that many more gravity measurements were needed, there is no indication that he anticipated the complex pattern of gravity anomalies that would be defined as more data were obtained. For example, he probably did not anticipate that within 15 mi of the Salt Lake City station Bouguer anomaly values 45 mgal higher or 26 mgal lower would be obtained. He certainly did not surmise that the 3 stations in Colorado west of the plains had an average isostatic anomaly of +29 mgal, whereas the next 8 stations established in this region would have average isostatic anomalies of -3 mgal, or that the average isostatic anomaly for his 10 "standard" stations was about 10 mgal lower than the average of the midcontinent region. The problem presented by the fact that the gravity observations were not representative of the areas he was comparing was compounded by the size of the area he selected to obtain average elevations around the gravity stations. Gilbert's 30-mi radius for the area over which he averaged the elevations is valid for the Great Basin but assumes a much more local compensation than exists in the Rocky Mountains and the Great Plains. His mean-plain anomaly values for Pikes Peak and Gunnison were +0.078 and +0.072 dyn, whereas Putnam, using a larger area and an absolute reference, computed values of +0.006 and +0.035 dyn. Thus, Putnam's computations using the same data suggest considerable compensation for the Rocky Mountains and do not support Gilbert's conclusion: "The whole Rocky Mountain plateau, regarded as a prominence on a broader plateau, is sustained by the rigidity of the lithosphere" (Gilbert, 1895a, p. 70). Nevertheless, his broader conclusion: "Nearly all the local peculiarities of gravity admit of simple and rational explanation on the theory that the continent as a whole is approximately isostatic. . . " (Gilbert, 1895a, p. 73) is correct.

Gilbert (1895b) published another paper in the *Journal of Geology* on the gravity data: "New Light on Isostasy." He was again led by the inadequate data and his method of reducing the data to conclusions that were partly invalid. He wrote:

In the Rocky Mountains of Colorado there are two stations, at Pikes Peak and Gunnison, and the excess of gravity determined at these stations is equivalent to the attraction of a rock layer 2200 feet thick. . . .

---

[2]Gilbert and Putnam both expressed the amplitude of gravity anomalies in dynes, a unit of force, rather than Gals, a unit of acceleration. Because they indicate the force of attraction on a mass of 1 g, the force in dynes is numerically equal to the acceleration in Gals.

These results tend to show that the earth is able to bear on its surface greater loads than American geologists, myself included, have been disposed to admit. They indicate that unloading and loading through degradation and deposition cannot be the cause of continued rise of mountain ridges with reference to adjacent valleys, but that, on the contrary, the rising of the mountain ridges, or orogenic corrugation, is directly opposed by gravity and is accomplished by independent forces in spite of gravitation resistance.

While the new data thus indicate that the law of isostasy does not obtain in the case of single ridges of the size of a large mountain range they agree with all other systems of gravity measurements in declaring the isostasy of the greater features of relief. [Gilbert, 1895b, p. 333]

Gilbert had, as had Putnam with the same data, established that in the United States isostasy was operating on a regional scale, but had reached the wrong conclusion concerning the isostatic equilibrium of the Rocky Mountains. He had estimated the gravity anomalies at each station caused by the difference in density of the rocks underlying the station from the average density of the crust. Apparently, independently of Putnam, he applied a technique for correcting the gravity values for the difference between the elevation of the gravity station and the regional elevation that facilitated studies of the departure from isostatic equilibrium. The techniques that he used to compute residuals and reduction to the plain suggest that he was not following the lead taken by others who had worked with gravity data but was developing nearly each step on his own. Although geophysicists and geodesists have not generally adopted Gilbert's methods of reducing gravity data, the method is sound. Gilbert's problems arose from an inadequate number of observations and the selection of too small an area for averaging elevations. He was probably influenced by his study of Lake Bonneville, where isostatic compensation is much more local than in areas to the east. The diameter of the circles over which he averaged elevations was approximately one-half the width of the Lake Bonneville basin. My own studies 70 yr later (Mabey, 1966) indicated that Gilbert's reduction methods are more applicable to the Great Basin than the more sophisticated isostatic reductions. The area of averaging he used is valid for the Great Basin and much better for this region than the larger area used by Putnam.

INTERPRETATION OF GRAVITY ANOMALIES

In 1894 the Coast and Geodetic Survey established 24 gravity stations, making a total of about 33 stations of good accuracy (± 10 mgal) in the United States. Gilbert hoped that the program of gravity measurements would be "continued for a few years with equal energy and skill" (Gilbert 1895a, p. 61), and he identified areas of investigation he considered particularly important. However, the production of gravity measurements in the next few years was low. Six were added in 1895, four in 1896, but none in 1897 and 1898. By the start of 1909, when an expanded program of gravity measurements was begun, a total of only 47 gravity stations had been established in the United States (Duerksen, 1949).

By 1913, 124 gravity stations had been established in the United States, and important papers on isostasy and gravity had been published. Hayford's studies (1909, 1910) on the deflection of the vertical, in which he divided the United States into 10 regions, led him to conclude "that while there are indications that the depth of compensation is greater in eastern and central portions of the United States than in the western portion, the evidence is not strong enough to prove that there is a real difference in the depth of

compensation in different regions" (Hayford, 1909, p. 143). Perhaps the evidence did not prove the difference, but the difference exists, and we must wonder why Hayford did not explore this possibility further when he and Bowie were using the gravity data to determine the depth of compensation. Whatever the reason, Hayford and Bowie (1912) assumed in their gravity studies that the depth of compensation was essentially uniform over the United States, and proceeded to attempt to force the gravity data from all regions to fit a single model. The model assumed a uniform depth of compensation, above which there was uniform vertical distribution of density and below which there were no horizontal variations in density. By using such a restricted model, they missed an opportunity to use the data to determine some important information on tectonic processes, and they were followed in this or similar simplified assumptions by most geodesists for many years. The geodesists were concerned primarily with the figure of the Earth and were attempting to minimize the complexities introduced into the gravity anomalies by local and regional geology. Perhaps they did not fully appreciate the geologic evidence that the crust of the Earth was behaving quite differently in different regions and that this difference was likely related to mass anomalies at depth.

In 1913 when Gilbert was 70 yr old, he once again examined the gravity data in the United States in what Davis (1926) said "was almost his last work on any subject. . . ." He wrote a paper "Interpretation of Anomalies of Gravity" (Gilbert, 1914), which was published as a chapter in a U.S. Geological Survey Professional Paper. The paper indicated that he was concerned about the treatment of gravity data by geodesists and was attempting to encourage a broader approach to the analysis of the data. He summarized the assumptions made by others in reducing the gravity data to the isostatic anomaly thus: "(1) that the compensation is perfect, (2) that the compensation defect or excess is uniformly distributed throughout the column, (3) that the depth of compensation, 122 kilometers, is not subject to variation from place to place" (Gilbert, 1914, p. 30). To these Gilbert added the implicit assumption that the Earth below the depth of compensation was composed of concentric homogeneous shells and thus did not contribute to gravity anomalies at the surface. Gilbert wrote:

The inexactness of the three assumptions is explicitly and fully recognized by the authors cited (Hayford and Bowie), but their discussion gives prominence only to the inexactness of the first. In a general way they interpret the anomalies as due to uncompensated local excesses or defects of mass in the crustal layer, or in other words, to imperfection of isostatic adjustment. The present paper will consider the possibilities of interpretation connected with the other specific assumption. [Gilbert, 1914, p. 30]

Gilbert computed the gravity effect of a modest anomaly in the vertical distribution of density above an assumed depth of compensation of 122 km and the effect of variation in the depth of compensation. Relative to the first assumption he concluded:

From this I infer that the anomalies may be in part due to irregularities in the vertical distribution of densities, or that such irregularities are competent, alike in the nature of their influence and in its possible amount, to cause such anomalies of gravity as have been discovered. [Gilbert, 1914, p. 31]

Concerning the second assumption, Gilbert wrote:

In view of the recognized heterogeneity of crustal material it appears to be both possible and probable that the depth at which material is sufficiently mobile to effect isostatic adjustment is subject not only to regional but to highly localized variation.

But he concluded:

I am disposed to regard variation in depth of compensation as decidedly less available in interpreting anomalies than variation in the vertical distributions of densities. [Gilbert, 1914, p. 31, 32]

Gilbert reproduced the gravity map of the United States and examined the major regional anomalies and their relation to geologic provinces. He concluded:

The correspondences are so slight that they may be regarded as accidental, and the general relation is that of independence and discordance. The result is of limited significance because the visible structure elements may constitute but a small fraction of the structure of the crust but so far as it goes it fails to support the hypothesis that the anomalies of gravity are due exclusively to anomalies in the vertical distribution of density within the crust. . . . [Gilbert, 1914, p. 34]

In the 1914 paper, Gilbert indicated that he thought the gravity data "show that the isostatic adjustment of the earth's crust is nearly perfect" (Gilbert, 1914, p. 29). Exactly what he meant by this is not clear. Earlier he had argued that large loads could be supported by the strength of the crust, and I conclude that his concept of perfect adjustment applied to broad geologic features and not to local features such as the ranges in the Basin and Range province. At that time no detailed gravity surveys of local geologic features had been made, and Gilbert's report does not indicate that he had considered the usefulness of gravity data in the study of local geology.

The sections of the paper on "The Locus of Adjustment" and "Interpretation by Nucleal Heterogeneity" provide an insight to Gilbert's thinking of tectonics. He cited as geologic evidence of mobility somewhere below the surface the crustal shortening in overthrust belts, the crustal extension of the Basin and Range province, and the eruption of liquid rock at the surface. He wrote:

The conception thus engendered, of a relatively mobile layer separating a less mobile layer above from a nearly immobile nucleus, appears to me in full accord with the evidence which geodesy affords of isostatic adjustment. The geodetic "depth of compensation" agrees with such suggestions as to the position of the horizon of maximum mobility as might be afforded by the volcanic and diastrophic phenomena. The existence of a horizon of mobility accords with the inference of approximate perfection of isostatic adjustment.

It is not necessary to suppose that the degree of mobility at the horizon of mobility is that of a liquid at the surface. When such mobility is attained by any but the densest rocks eruption takes place. It is not necessary to think of the degree of mobility as uniform, either from place to place or from time to time. Its place variation would naturally be coordinate with that of rock types, and its time variation coordinate with epochs of elevation and subsidence. Neither should the depth of the horizon of maximum mobility be thought of as uniform . . . .

It is my own view that the inner part of the nucleus is not merely hot, but very hot. . . . An enormous temperature implies an enormous store of heat. This is the source of the energy involved in the hypogene activities of the earth, and it is fed to the crustal region by conduction. . . . The deformations which have not only developed but perpetually remodeled the continents

have a source below the surface. Their method probably involved reactions between temperature, pressure, and the physico-chemic constitution of rocks, but these reactions, like the superficial reactions, yield running-down processes and do not afford a fundamental explanation of crustal activity. The factor to which I appeal is primordial heterogeneity of earth material, a heterogeneity which gives diversity to the flow of heat energy and to the physical and chemic changes of crustal regions. It does not seem sufficient that the crust be heterogeneous; there should be heterogeneity also below the horizon of mobility. [Gilbert, 1914, p. 35]

Gilbert's concluding paragraph was:

Starting with geodetic and topographic data and assuming certain uniformities, Hayford and Bowie have demonstrated isostasy and developed a gravity anomaly map. The map contains a body of observational data coordinate with the geodetic and topographic. By a future mathematical discussion which treats the three bodies of data together and which recognizes alternative interpretations of the anomalies of gravity it may be possible to practically demonstrate the meaning of the anomaly map. At present the map seems to express chiefly an effect of heterogeneity in the nucleus and an effect of irregularity in the vertical distribution of densities within the crust. [Gilbert, 1914, p. 37]

CONCLUSIONS

What has been the impact of Gilbert's work on the development of the use of gravity data and the study of isostasy? The published record suggests surprisingly little. Although he identified the beautiful research opportunity presented to the geophysicist by the Lake Bonneville data, the challenge was not answered for 70 yr. Because of his background in geology, Gilbert understood the relationship between gravity anomalies and isostasy better than the early geodesists. If Gilbert's lead had been followed by the early workers with gravity data, the contribution of gravity data to the study of tectonic processes would have been much greater. But geodesists largely ignored his work. For example, Heiskanen and Vening Meinesz (1958), writing on the historical development of the idea of isostasy, do not mention Gilbert.

How can this lack of impact be explained? Partly by Gilbert's limited "mathematical resources." He developed no new formula nor was his quanitative analysis of the data on a level to inspire most geophysicists. In the Lake Bonneville study, he involved Woodward to provide some needed computations, but in later years he apparently relied on his own skills in this area. And aside from his Lake Bonneville study, he produced little new basic data on gravity and isostasy. Perhaps in Gilbert's work we observe the development of a problem that plagued earth science in the United States through the first half of the twentieth century and is only now being resolved. Geology, which up to Gilbert's time was almost entirely observational, was moving into the era when the solution of important problems required skill in physics and mathematics not generally possessed by the geologist. Problems requiring these skills were left to geodesists and seismologists who, while possessing the skill in mathematics and physics, commonly lacked an in-depth understanding of the geology. Often, as with the geodesists and seismologists in the U.S. Coast and Geodetic Survey and the geologists in the U.S. Geological Survey, groups that should have been working were separated by organizational barriers that increased communication problems that existed because of the different backgrounds. However, at least one geodesist found Gilbert personally inspiring. Three days before

Gilbert died, J. F. Hayford wrote in a letter intended to be presented to Gilbert on his 75th birthday:

Soon after I came to Washington as a young man fresh from college a certain few men came to dominate my ways of thinking, — you were among them. . . . Your early discernment of some of the relations between geology and geodesy, — and of the importance of those relations, — was largely influential in giving me the courage to persist in my groping for the truth in regard to isostasy at a period when I had but glimmering glimpses of it.

Gilbert's three ventures into gravity and isostasy were separated by periods of several years, and little or no continuity exists between them. The conclusions reached are not mutually consistent, and nowhere does Gilbert examine the inconsistencies. His final paper does not contain a single reference to his earlier works. He does not explain how his conclusions that the strength of the crust could support the Rocky Mountains made in 1895 related to the conclusions in 1890, that the strength of the crust could not support the much smaller load of Lake Bonneville, or how either of these conclusions relate to the concept of perfect isostasy advocated in 1914. The reports of his studies do not build one upon another toward a unified theory of isostasy. Rather, Gilbert appears to be hypothesizing to the end and encouraging the theorists to do the same. His objective apparently was not to develop and argue for a theory but to point out that the specialists needed to take a broader look at the problem. In that, he was correct.

Gilbert (1914, p. 35) wrote:

The inner earth is the inalienable playground of the imagination. Once it contained the forges of blacksmith gods; or it was the birthplace of our race, or the home or prison of disembodied spirits. Later Symmes hollowed from it a vast habitable empire, concave like the world of Koresh. Science now claims exclusive title but holds it chiefly for speculative purposes; and the freedom of speculation practically recognizes but two limitations: The inner earth is dense, and it is rigid. As to all other properties opinion is untrammeled.

Gilbert speculated on the inner Earth and demonstrated remarkable insight into the meaning of the geological and geophysical information available to him. His vision as to what lines of research were likely to prove fruitful is impressive.

Gilbert established with the Lake Bonneville data that the strength of the crust could not support the load of water as the basin filled; it subsided, but recovered when the load was removed. He recognized that this was strong evidence that isostasy was a reality. He also correctly concluded that some large loads could be supported by the crust. He knew that the Lake Bonneville shorelines provide an outstanding opportunity for a study of dynamic behavior of the crust.

Using the first gravity data obtained in the United States, he identified some of the major uses to which a more extensive data set could be put and found that the few stations available to him confirmed that the major features of the topography of the United States were in approximate isostatic equilibrium. When substantially more data were available 20 yr later, Gilbert used an understanding of geologic processes in an attempt to direct the geodesists toward a broader approach to their analysis of the data. His own conclusions that the major deformations observed in the crust reflect "horizontal movements of the upper rocks without corresponding movements of the nucleus and thereby imply mobility in an intervening layer" and his identification of the

driving force as a "primordial heterogeneity of earth material, a heterogeneity which gives diversity to the flow of heat energy" are remarkably compatible with current theories.

REFERENCES CITED

Airy, G. B., 1885, On the computations of the effect of the attraction of the mountain masses as disturbing the apparent astronomical latitude of stations in geodetic surveys: Transactions of the Royal Society of London, ser. B, v. 145.

Boscovich, R. J., 1750, De litteraria expeditione per pontificiam ditionem: 475 p.

Bouguer, Pierre, 1749, La figure de la terre: Paris, 364 p.

Cook, K. L., and others, 1975, Simple Bouguer gravity anomaly map of Utah: Utah Geological and Mineralogical Survey Map 37.

Crittenden, M. D., Jr., 1963a, Effective viscosity of the earth derived from isostatic loading of Pleistocene Lake Bonneville: Journal of Geophysical Research, v. 68, p. 5517–5530.

———1963b, New data on the isostatic deformation of Lake Bonneville: U.S. Geological Survey Professional Paper 454-E, 31 p.

Davis, W. M., 1926, Biographical memoir of Grove Karl Gilbert, 1843–1918: National Academy of Sciences Memoir, v. 21, 5th Memoir, 303 p.

Delaney, John P., 1940, Leonardo da Vinci on isostasy: Science, v. 91, p. 546.

Duerksen, J. A., 1949, Pendulum gravity data in the United States: U.S. Coast and Geodetic Survey Special Publication no. 244, 217 p.

Dutton, C. E., 1889, On some of the greater problems of physical geology: Philosophical Society of Washington Bulletin, v. 11, p. 51–64.

Gilbert, G. K., 1886, The inculcation of scientific method by example: American Journal of Science, v. 31, p. 284–299.

———1890a, Lake Bonneville: U.S. Geological Survey Monograph 1, 438 p.

———1890b, The strength of the earth's crust [abs. with discussion]: Geological Society of America Bulletin, v. 1, p. 23–27.

———1892, Remarks on paper "On some of the greater problems of physical geology" [abs.]: Philosophical Society of Washington, v. 11, p. i–xxxi; 536–537.

———1894, A report on a geologic examination of some Coast and Geodetic Survey gravity stations, in Report of the Superintendent of the U.S. Coast and Geodetic Survey showing the progress of the work during fiscal year ending with June, 1894: U.S. Coast and Geodetic Survey, Part II, App. 1, p. 51-55.

———1895a, Notes on the gravity determinations reported by Mr. G. R. Putnam: Philosophical Society of Washington, v. 13, p. 61–75.

———1895b, New light on isostasy: Journal of Geology, v. 3, p. 331–334.

———1914, Interpretation of anomalies of gravity: U.S. Geological Survey Professional Paper 85-C, p. 29–37.

Hayford, J. F., 1909, The figure of the earth and isostasy from measurements in the United States: U.S. Coast and Geodetic Survey, 178 p.

———1910, Supplemental of investigations in 1909 of the figure of the earth and isostasy: U.S. Coast and Geodetic Survey, 80 p.

Hayford, J. F., and Bowie, William, 1912, The effect of topography and isostatic compensation upon the intensity of gravity: U.S. Coast and Geodetic Survey Special Publication no. 10, 80 p.

Heiskanen, W. A., and Vening Meinesz, F. A., 1958, The earth and its gravity field: McGraw-Hill, 470 p.

Mabey, D. R., 1966, Relation between Bouguer gravity anomalies and regional topography is Nevada and the eastern Snake River Plain, Idaho: U.S. Geological Survey Professional Paper 550-B, p. 108–110.

Pratt, J. G., 1855, On the attraction of the Himalaya Mountains and the

68 D. R. MABEY

elevated regions beyond upon the plumb-line in India: Transactions of the Royal Society of London, ser. B, v. 145.

Putnam, G. R., 1894, Relative determinations of gravity with half-second pendulums, and other pendulum investigations, *in* Report of the Superintendent of the U.S. Coast and Geodetic Survey showing the progress of the work during the fiscal year ending with June 1894: U.S. Coast and Geodetic Survey, Part II, App. 1, p. 9–50.

Walcott, R. I., 1970, Flexural rigidity, thickness and viscosity of the lithosphere: Journal of Geophysical Research, v. 75, p. 3941–3954.

MANUSCRIPT RECEIVED BY THE SOCIETY AUGUST 27, 1979
MANUSCRIPT ACCEPTED MAY 20, 1980

Geological Society of America
Special Paper 183
1980

Gilbert and the Moon

FAROUK EL-BAZ

National Air and Space Museum, Smithsonian Institution, Washington, D.C. 20560

> *If the record of her scarred face has now been read aright, all that remains*
> *of the old narrative is its denouement: the moon is dead.*
>
> G. K. Gilbert

ABSTRACT

Sketches made by G. K. Gilbert and based on telescopic observations of the Moon look amazingly similar to photographs obtained 75 yr later by spacecraft. He was very successful in correlating lunar surface features with counterparts on Earth. His observations and experiments led him to the conclusion that most lunar craters are the product of impact. After establishing this, he studied the Coon (Meteor) Crater of Arizona. He did not have as much success applying what he had learned from the Moon to the terrestrial case. He conducted a topographic study of the crater to check whether there was an added volume due to the incoming projectile. An overestimation of the size of the meteorite and neglect of the possibility of its fusion, evaporation, and ejection forced him to rule out an impact origin for this crater.

In his observations on lunar features, Gilbert had expressed the basic elements of a lunar stratigraphic system. His discussion of crater rays, and particularly of the "sculpture" that surrounds the Imbrium basin, greatly influenced the thinking of lunar geologists of our day. Coupled with his recognition of the importance of crater density and overlap relationships, he can be easily considered the father of lunar stratigraphy. Today there is a crater on the Moon bearing the name of Gilbert in commemoration of his many contributions to geology.

INTRODUCTION

In attempting to understand the surface features of the Moon, geologists transported their controversial concepts 400,000 km away. Before the space age, evidence for the origin of lunar surface features was limited to telescopic observations and comparisons with terrestrial features. Pivotal to the evaluation of such evidence was consideration of the concepts of catastrophism and uniformitarianism. According to the former concept, the surface of the Earth is thought to have been formed by several catastrophic events. According to the latter, geologic features are viewed as the result of a gradually evolving Earth.

At the end of the nineteenth century, uniformitarianism had gained many supporters. Thus, the Earth was viewed as a product of slow and continuous change that could be easily explained in terms of the same processes that are observable today. The Moon was considered along similar lines of thought, with the consequence that most of its surface features were interpreted to be the result of endogenetic processes. Gradual and prolonged volcanism was credited with the formation of the lunar surface features, particularly its craters, which constitute the most predominant of lunar landforms.

LUNAR CRATERS

Gilbert was fully aware that most writers of his time, and before him, assumed that lunar craters were of volcanic origin. Because these craters differed in size, abundance, and shape from terrestrial volcanoes, they were thought to represent a special type of volcanism that resulted from the Moon's peculiar physical conditions.

He realized that the problem was largely one of interpretation of observed forms. His observations were limited to 18 nights during two lunations in August through October of 1892. He used mostly the 400 power, 67.31-cm (26.5-in.) refracting telescope of the U.S. Naval Observatory.

Gilbert recognized that although there existed a great range of lunar crater sizes, within this range were varieties whose occurrence more or less correlated with size. He further observed that their intergradation was so perfect that they could easily be regarded as phases of a single type.

With this in mind, Gilbert started to examine the detailed characteristics of the type form of lunar craters:

Picture to yourself a circular plain, ten, twenty, fifty, or one hundred miles in diameter, surrounded by an acclivity which everywhere rises steeply but irregularly to a rude terrace, above which is a circular cliff likewise facing inward toward the plain. This cliff is the inner face of a rugged, compound, angular ridge, composed of shorter ridges which overlap one another, but all trend concentrically. Seen from above, this ridge calls to mind a wreath, and it has been so named. From the outer edge of the wreath a gentle slope descends in all directions to the general surface of the moon, which it is convenient to call here the outer plain. The outer slope of the crater may be identical in surface character with the outer plain, or it may be radially and somewhat delicately ridged, as though by streams of lava. The inner slope, from the base of the cliff to the margin of the inner plain, is broken by uneven and discontinuous terraces, which have the peculiar habit of land-slip terraces as one sees them about the flanks of a plateau capped by a heavy sheet of basalt. From the center of the inner plain rises a hill or mountain, sometimes symmetric but usually irregular and crowned by several peaks. [Gilbert, 1893, p. 243–244]

This description of lunar crater forms is among the best ever published up to Gilbert's time. His drawing of an example so typifies lunar craters that it looks hauntingly similar to spacecraft views of the craters (Fig. 1). Furthermore, his cross section that is based on this drawing does not vary much from profiles drawn on

the basis of photogrammetric measurements of stereophotographs taken by the Apollo missions from lunar orbit (Fig. 2).

To Gilbert, the central hill was an important part of the lunar crater form, but it perplexed him that these hills were not universally present. This was the case even in craters of similar sizes. Also, in very large craters (150 km and larger), the occurrence of a central hill was rare; it disappeared altogether in the largest of craters.

Considering that he lacked the close-up photography we have today, his observations of crater morphology and its relation to size are amazingly accurate. He correctly recognized that the depth of a crater varies with the width, but less rapidly. Detailed measurements from Apollo photographs (Pike, 1974) have demonstrated that the following relationships apply, where R = crater depth and D = crater diameter: $R = 0.196D^{1.010}$ for small fresh

Figure 1. Top, sketch of the type form of lunar crater (Gilbert, 1893, p. 243). Bottom, oblique view of the lunar crater Theophilus, 100 km in diameter, obtained by Lunar Orbiter III (frame 78M).

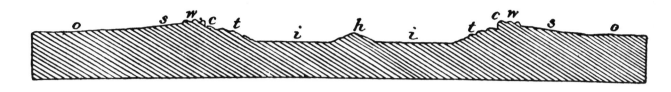

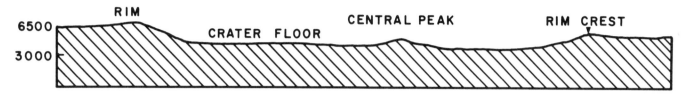

Figure 2. Top, cross profile of lunar crater: o, outer plain; s, outer slope; w, wreath; c, inner cliff; t, terraced inner slope; i, inner plain; h, central hill (Gilbert, 1893, p. 244). Bottom, profile of the lunar crater Plinius, 43 km in diameter, based on Lunar Topographic Orthophotomap 60 B1 at 1:250,000 scale. Vertical exaggeration is approximately 0.6X.

craters (less than about 15 km) and $R = 1.044D^{0.301}$ for larger fresh craters.

What he called the inner plain is now referred to as the crater floor. Gilbert noted that this inner plain was a constant feature in large- and medium-sized craters, but it was rare in craters with small diameters: "to my eye the interiors of most craters under four miles and of all under two miles appear as simple cups" (Gilbert, 1893, p. 245). Present data place the diameter of transition between simple and complex craters at about 10 to 20 km (Pike, 1975). Gilbert had anticipated this by noting the absence of central peaks as crater diameter decreased.

Considering the limitation of his instrument:

I am conscious that as the limit of telescopic vision is approached, the details of craters must disappear before the craters themselves are lost, and I am therefore anxious to have this observation verified by those who are able to use higher powers than I could. [Gilbert, 1893, p. 245]

As it turned out, the observations of Gilbert were found to hold, even after the limit of resolution was increased several fold by the use of the Apollo panoramic camera, which has a lens with a 610-mm focal length and was flown only 100 km above the lunar terrain (Masursky and others, 1978).

As he completed the description of the form of lunar craters, Gilbert had in effect provided us with a classification of their morphologies. His illustration of their variety includes most of the basic forms of lunar craters as photographed by spacecraft (Fig. 3). This thorough morphological analysis allowed him to contrast lunar craters with terrestrial counterparts in order to deduce their genesis. Naturally, most terrestrial craters known at his time were volcanic in origin.

THE IMPACT THEORY

Perhaps Gilbert's most significant contribution to lunar studies was his 1893 discussion of the origin of lunar craters. He presented a convincing case that these features are the product of impact. The theory was developed by pointing out the differences in form between the majority of terrestrial volcanoes and lunar craters:

Ninety-nine times in one hundred the bottom of the lunar crater lies lower than the outer plain; ninety-nine times in a hundred the bottom of the Vesuvian crater lies higher than the outer plain. Ordinarily the inner height

of the lunar crater rim is more than double its outer height; ordinarily the outer height of the Vesuvian crater rim is more than double its inner height. The lunar crater is sunk in the lunar plain; the Vesuvian is perched on a mountain top. The rim of the Vesuvian crater is not developed, like the lunar, into a complex wreath, but slopes outward and inward from a simple crest-line. If the Vesuvian crater has a central hill, that hill bears a crater at summit and is a miniature reproduction of the outer cone; the central hill of the lunar crater is entire, and is distinct in topographic character from the circling rim. The inner cone of a Vesuvian volcano may rise far higher than the outer; the central hill of the lunar crater never rises to the height of the rim and rarely to the level of the outer plain. The smooth inner plain characteristic of so many lunar craters is either rare or unknown in craters of Vesuvian type. Thus, through the expression of every feature the lunar crater emphatically denies kinship with the ordinary volcanoes on the earth. [Gilbert, 1893, p. 250]

Gilbert noted that terrestrial maars do resemble small lunar craters but provide no explanation for intermediate and larger ones. He also examined and rejected other proposed hypotheses such as the "snow theory," which stated that water vapor derived from localized pools fell as snow upon the pools and surrounding areas, eventually forming circular ridges.

He concluded that "all other theories which I have been able to discover appeal in one way or another to the collision of other bodies with the moon's surface" (Gilbert, 1893, p. 256). For the impact theory of lunar crater formation he coined the term "meteoric." He then proceeded to make experiments to study the shape of craters resulting from impact (Fig. 4). He used simple analogies:

If a pebble be dropped into a pool of pasty mud, if a rain drop falls upon the slimy surface of a sea marsh when the tide is low, or if any projectile be made to strike any plastic body with suitable velocity, the scar produced by the impact has the form of a crater. This crater has a raised rim suggestive of the wreath of lunar craters. With proper adjustment of material, size of projectile, and velocity of the impact, such a crater scar may be made to have a central hill. [Gilbert, 1893, p. 256]

His conclusion that a "meteoric" theory best explained the observed lunar features was coupled with an understanding that it was not without problems. The most significant of these problems to him was the general circularity of the craters, which he felt indicated predominantly vertical impacts. To explain this observation he proposed that the Moon formed from the coalescing of

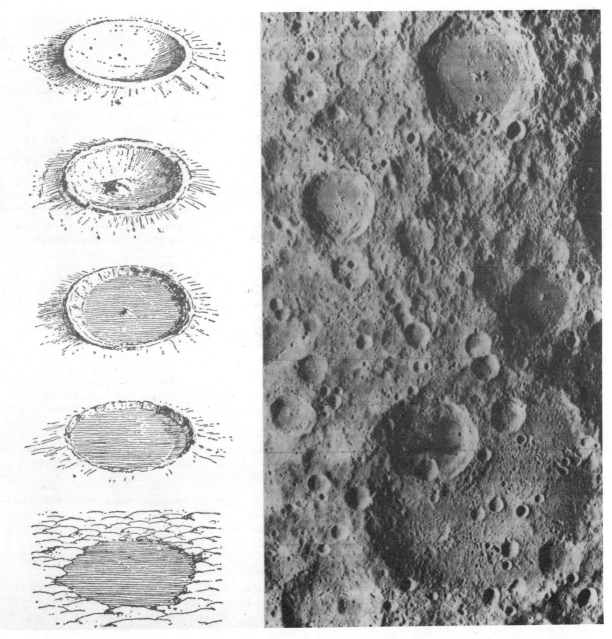

Figure 3. Left, varieties of lunar craters as related to size. The uppermost sketch represents the form of the smallest craters, the lowermost, the largest (Gilbert, 1893, p. 245). Right, part of the highly cratered highlands on the lunar far side illustrating that all basic crater forms have counterparts in Gilbert's sketches. The largest crater in lower right is Gagarin (Lunar Orbiter I frame 116M).

debris located in a ring around the Earth and that craters were the scars of the final stages of collision. Recent experimental labora- tory data, however, show that oblique impacts may produce circular craters (Gault, 1975).

METEOR CRATER

Now that Gilbert was satisfied with the "meteoric" origin of most lunar craters, he turned to the Earth in search of a convincing analogue. Ironically, he did not have as much success applying to terrestrial cases what he had learned from the Moon. In a study of Meteor Crater in northern Arizona (then called Coon Butte or Coon Crater, later also called Barringer Crater), he concluded that it resulted from a steam explosion (Gilbert, 1896).

This crater is more than 1 km in diameter and nearly 200 m deep (Fig. 5). It is located in a region of flat-lying Permian and Triassic sedimentary rocks. Although the structure stands out as a striking anomaly on a flat plateau, had it occurred 30 km to the northwest, it would have been nestled inconspicuously among Quaternary volcanic craters (Mutch, 1972, p. 84).

Gilbert correctly noted that

Figure 4. Experiments illustrating the formation of craters by impact. Scales are in inches. Top, balls of clay the size of the one shown in the picture were thrown against a slab of clay. The difference in result depended on difference in velocity of impact (G. K. Gilbert photo 842, about 1891). Bottom, three balls of clay like the one pictured were thrown obliquely against a target of clay (G. K. Gilbert 843, about 1891).

The crater differs from the ordinary volcanic structure of that name in that it contains no volcanic rock. The circling sides of the bowl show limestone and sandstone, and the rim is wholly composed of these materials. On the slopes of this crater and on the plain around about many pieces of iron have been found, not iron ore, but the metal itself, and this substance is foreign to the limestone of the plain and to all other formations of the region. [Gilbert, 1896, p. 3]

Thus, Gilbert's field observations (Fig. 6) allowed him to recognize all the basic characteristics of this crater including the unusual features, which required explanation: (1) the crater was composed of nonvolcanic materials and (2) in association with the crater were scattered iron masses, which were foreign to the local materials.

With what appeared as reasonable assumptions, he started searching for the most suitable theory of origin:

If the crater was produced by the collision and penetration of a stellar body that body now lay beneath the bowl, but not so if the crater resulted from explosion. Any observation which would determine the presence or absence of a buried star might therefore serve as a crucial test. [Gilbert, 1896, p. 5]

To search for the "stellar" material by means of a shaft or drill hole was not considered feasible because of the expense involved. Therefore, Gilbert had to consider indirect methods. He proposed to solve the problem by topographic means:

If the crater were produced by explosion the material contained in the rim, being identical with that removed from the hollow, is of equal amount; but if a star entered the hole the hole was partly filled thereby, and the remaining hollow must be less in volume than the rim. The presence or absence of the star might therefore be tested by measuring the cubic contents of the hollow and of the rim and comparing the two. [Gilbert, 1896, p. 5]

Naturally, he could find no evidence of the presence of the impacting projectile, which he expected to be lodged below the crater floor. His basic error was in the overestimation of the size of the projectile and in his neglect of the possibility of its fusion, evaporation, and ejection. He expected to find local magnetic variations and a discrepancy between the volume of ejecta and volume of the cavity due to the presence of the projectile.

Thus, Gilbert felt forced to rule out an impact origin for the crater. Today all workers are convinced that it was formed by a meteor impact because of abundant evidence of disturbance and shock metamorphism: fracturing, brecciation, presence of glass formed by the shock melting, and occurrences of two high-pressure forms of silica: coesite (Chao and others, 1960) and stishovite (Chao and others, 1962). Before their discovery at Meteor Crater, these two minerals were known only from laboratory synthesis (Mutch, 1972, p. 86).

Today's estimate of the size of the impacting body is much smaller than what was assumed by Gilbert. The original body is now assumed to have been about 25 m in diameter and to have had a mass of about 65,000 metric tons. The impact is estimated to have released energy equivalent to nearly 2×10^6 metric tons of TNT (French, 1977, p. 151).

Although Gilbert felt compelled to reject the impact origin for this crater, his ever-present objectivity led him in the end to the correct conclusion that all the evidence was not in yet. He was familiar with the ideas, put forth by Edwin E. Howell, that the projectile may have consisted of small iron chunks embedded in a nonmagnetic material or that the target area may have been compressed by the impact. These theories caused Gilbert to doubt his own conclusions:

These considerations are eminently pertinent to the study of the crater. . . but the fact which is particularly worthy of note at the present time is their ability to unsettle a conclusion that was beginning to feel itself secure. This illustrates the tentative nature, not only of the hypotheses of Science, but of what Science calls its results. . . However grand, however widely accepted, however useful its conclusion, none is so sure that it can not be called in question by a newly discovered fact. In the domain of the world's knowledge there is no infallibility. [Gilbert, 1896, p. 12]

OTHER LUNAR FEATURES

Gilbert also made detailed observations of features other than craters. Particularly noteworthy is his description of the sinuous

Figure 5. Aerial view of Meteor Crater, Arizona (courtesy of David J. Roddy, U.S. Geological Survey).

rises on the lunar surface known as mare ridges. He noted that they have anticlinal and monoclinal forms, but are so gentle of slope that they are seen only near the terminator. Today, some of our best photographs of these ridges were indeed taken near the terminator (Fig. 7). Some ridges mark the boundaries of raised shelves in the maria, and it was probably these features he described as monoclinal. This interpretation was not confirmed until in 1972 the Apollo Lunar Sounder provided us with profiles across lunar basins with circular ridge systems such as those in Mare Serenitatis (Maxwell and others, 1975).

He also described rilles, rille pits, and large furrows that he observed telescopically. However, none of the features that he described had as much significance to lunar geology as the crater rays and the "sculpture" that surrounds the Imbrium impact basin.

Gilbert called the rays "white streaks," which he described as

bands of color, sometimes faint, sometimes brilliant, but always indefinitely outlined, like the tail of a comet. . . . Some of them stretch for long distances across the moon's surface. . . . They pass up and down the slopes of craters without either modifying their forms or being interrupted by them. The more prominent of them, and probably all, occur in systems, and those of each system radiate from some crater. [Gilbert, 1893, p. 284]

Gilbert was aware of the unpublished suggestion of a Mr. William Würdemann of Washington that a meteorite striking the Moon's surface with great force would spatter the "whitish matter" in various directions. The explanation appealed to Gilbert because it accounted for

the straightness of the rays, for their vanishing edges and ends, for their independence of topography, for their relation to craters, for the whiteness of the associated craters, and for the nimbus in which the rays sometimes unite close to the crater. [Gilbert, 1893, p. 285]

The idea was appealing, and a reasonable theory was established. However, because of his meticulous observations and keen mind, Gilbert went on to pose several questions that until recently continued to puzzle lunar workers:

What is the white substance? Why do its traces become faint in passing from the bright uplands to the dark plains? Why do wavy lines replace straight ones in the radiation from Copernicus? Why do certain great rays of Tycho's system trend toward a point of the rim and not toward the center of the crater? Why are several craters, especially Tycho, surrounded by a relatively dark band inside the bright nimbus? [Gilbert, 1893, p. 285-286]

Figure 6a. Photographs taken in 1891 by G. K. Gilbert of the interior of Coon Butte (Meteor Crater), Coconino County, Arizona. Top, view from the rim (photo 784). Bottom, view of the inner slope (photo 775).

It was only after detailed lunar photographs were obtained and samples were returned that adequate answers to some of these questions were found (El-Baz, 1975; Masursky and others, 1978).

Gilbert did even better observing, describing, and interpreting the features which he termed "sculpture." He observed that

the rims of certain craters are traversed by grooves or furrows, which arrest attention as exceptions to the general configuration. In the same

neighborhood such furrows exhibit parallelism of direction. Similar furrows appear as tracts between craters. . . . Tracing out these sculptured areas and plotting trend lines on a chart of the moon, I was soon able to recognize a system in their arrangement, and this led to the detection of fainter evidence of sculpture in yet other tracts. The trend lines converge toward a point near the middle of the plain called Mare Imbrium. [Gilbert, 1893, p. 275]

Thus, he was the first to recognize the extensive system of lineations and grooves known as Imbrium sculpture (Fig. 8). He

Figure 6b. Photographs taken in 1891 by G. K. Gilbert of the exterior of Coon Butte (Meteor Crater). Top, morainelike hills on outer slope (photo 777). Bottom, limestone block ejected from the crater (photo 754).

attributed this pattern to "a collision of exceptional importance" whose result was "the violent dispersion in all directions of a deluge of material—solid, pasty, and liquid." He was able to discriminate between "antediluvial" and "postdiluvial" features and went on to state that "by the outrush from the Mare Imbrium were introduced the elements necessary to a broad classification of the lunar surface" (Gilbert, 1893, p. 279).

LUNAR STRATIGRAPHY

In his address as Retiring President of the Philosophical Society of Washington, delivered December 10, 1892, Gilbert's first remarks were "the face which the moon turns ever toward us is a territory as large as North America, and, on the whole, it is probably better mapped" (Gilbert, 1893, p. 241). To this day, the

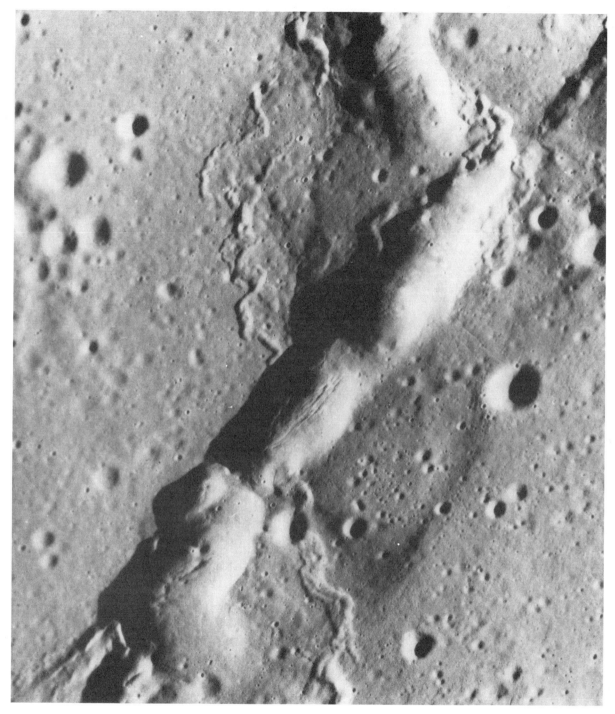

Figure 7. Lunar mare (wrinkle) ridge in the eastern part of Mare Serenitatis showing the effect of low Sun illumination angles in emphasizing the form of the 1-km-wide ridge (Apollo 17 panoramic camera photo 2313).

statement holds true, for the Moon has been mapped by more sophisticated systems than much of North America, not to mention the rest of the world (El-Baz and Ondrejka, 1978, p. 704). Lunar mapping, however, is the domain of selenographers, and interpretation is the realm of selenologists.

Gilbert tackled the questions concerning the origin of the Moon's features in much the same way that he approached any terrestrial geomorphologic problem. He recognized that the basic geologic principles would be the same, and he painstakingly examined the available hypotheses. He was modest enough to state that his comments were not based on "protracted observation nor on protracted study." If his observations were indeed less than thorough, it is not obvious from his presentation. In fact, he anticipated the lunar stratigraphic system in use today.

Figure 8. The Imbrium sculpture is seen here as grooves in the cratered highland material in the central part of the near side of the Moon (Apollo 15 metric camera photo 1411).

The basic elements of lunar stratigraphy can all be found in Gilbert's observations, including the recognition that (1) the Moon's surface is a continuous panorama of impact physiography; (2) the abundance of craters on the Moon's surface varies from place to place with fewer large craters in the maria; (3) the increase in crater size is accompanied by atrophy or degradation of crater form; (4) because craters overlap and the overlapping is never reciprocal, clear-cut age interpretations are possible; (5) the "white streaks" are crater rays; and, most significantly, (6) the "sculpture" on the near side of the Moon is due to the formation by impact of the Imbrium crater.

With the establishment of these basic ideas, Gilbert correctly recognized that "the Moon is dead" and that impact is the main surface-sculpting mechanism, which is also responsible for lateral movement of lunar surface materials. Study of the superposition of impact and other materials can, therefore, help in understanding the history of the lunar surface.

When the United States started preparing for the Apollo program, a major effort of geologic mapping of the Moon was initiated by the U.S. Geological Survey. Setting the stage were the papers by Shoemaker in 1960 and 1962 dealing with Meteor Crater and lunar craters, respectively. At the same time, Snoemaker and Hackman published the first stratigraphic discussion of the Moon; they were fully aware of the basic parameters that Gilbert established 80 yr earlier. Starting in the area of the crater Copernicus, they recognized that (1) it is younger than the neighboring Eratosthenes because its form is sharper and its rays are obvious and (2) Eratosthenes is younger than Imbrium because its ejecta overlap the Imbrium basin materials.

In all, Shoemaker and Hackman (1962) recognized five overlapping sets of deposits that, except for the first, they called systems: (from oldest to youngest) pre-Imbrian, Imbrian, Procellarian, Eratosthenian, and Copernican. This stratigraphy was applied to nearly the whole Earth-facing lunar hemisphere, based mainly on telescopic observations (Wilhelms, 1970; Wilhelms and McCauley, 1971).

After the acquisition of photographs of the far side of the Moon, it was recognized that an extension of the nearside stratigraphy could be employed on the far side as well (El-Baz, 1974, 1975; Wilhelms and El-Baz, 1977). At the present time, a Moon-wide time-stratigraphic sequence exists, as described below in order of decreasing relative age:

Pre-Nectarian. All materials formed before the Nectaris basin (a large impact basin on the Moon's near side that is older than Imbrium) and as far back as the formation of the Moon are classed as pre-Nectarian. The majority of pre-Nectarian units are distinguished on the lunar far side, including mostly materials of very old and subdued basins.

Nectarian System. This system includes all materials stratigraphically above and including Nectaris basin materials, up to and not including Imbrium basin strata (Stuart-Alexander and Wilhelms, 1974). Ejecta of the Nectaris basin that can be traced near the east-limb region allow recognition of these materials as a stratigraphic datum for the farside highlands.

Imbrian System. A large part of the lunar surface is occupied by ejecta surrounding both the Imbrium and Orientale basins. These form the lower and middle parts of the Imbrian System, respectively. Two-thirds of the mare materials belong to this system, particularly in the eastern maria of Crisium, Fecunditatis, Tranquillitatis, Nectaris, the dark annulus of Serenitatis, as well as most mare occurrences on the lunar far side.

Eratosthenian System. This system includes materials of rayless craters such as Eratosthenes. Most of these are believed to have once displayed rays that are no longer visible because of mixing due to prolonged micrometeorite bombardment as well as solar radiation. The system also includes about one-third of the mare surface materials on the lunar near side, particularly in Oceanus Procellarum.

Copernican System. This is stratigraphically the highest and, hence, the youngest lunar time-scale unit. It includes materials of fresh-appearing, intermediate- to high-albedo, bright-rayed craters. The system also includes exposures of very high albedo material on inner walls of Copernican and older craters, as well as other scarps. Brightness in these cases in interpreted as the result of fresh exposure by mass wasting and downslope movement along relatively steep slopes.

From this, it is clear that Gilbert's ideas on the Moon greatly affected the evolution of a lunar stratigraphic column. It was thus a well-deserved honor that the International Astronomical Union (IAU) decided in 1964 to name a lunar crater after him (Fig. 9). The crater is 63 km in diameter and is centered at lat 3.5°S, long 76.5°E. However, it is ironic that his contributions to knowledge of the Moon were not specifically mentioned in the IAU official transaction, which read (Menzel and others, 1971, p. 183):

Gilbert, G. K. (1843-1918), USA geologist; research on intrusive igneous features; introduced ideas of erosion, glaciation and river development to geomorphology; concept of glaciation in formation of Great Lakes.

Figure 9. Lunar crater named after Gilbert to honor his contributions to geology (Lunar Orbiter IV, frame 178H1).

REFERENCES CITED

Chao, E. C. T., Shoemaker, E. M., and Madsen, B. M., 1960, First natural occurrence of coesite: Science, v. 132, p. 220-222.

Chao, E. C. T., and others, 1962, Stishovite, SiO2, a very high pressure new mineral from Meteor Crater, Arizona: Journal of Geophysical Research, v. 67, p. 419-421.

El-Baz, F., 1974, Surface geology of the Moon: Annual Reviews of Astronomy and Astrophysics, v. 12, p. 135-165.

——1975, The Moon after Apollo: Icarus, v. 25, p. 495-537.

El-Baz, F., and Ondrejka, R. J., 1978, Earth orbital photography by the Large Format Camera: Proceedings of the Twelfth International Symposium on Remote Sensing of Environment, v. 1, p. 703-718.

French, B. M., 1977, The Moon book: New York, Penguin Books, 287 p.

Gault, D. E., 1975, Impact cratering, in Greeley, R., and Schultz, P., eds., A primer in lunar geology: Moffett Field, California, NASA Ames Research Center, p. 137-176.

Gilbert, G. K., 1893, The moon's face; A study of the origin of its features: Philosophical Society of Washington Bulletin, v. 12, p. 241–292.

——1896, The origin of hypotheses, illustrated by the discussion of a topographic problem: Science, v. 3, p. 1–12.

Masursky, H., Colton, G. W., and El-Baz, F., 1978, Apollo over the Moon: A view from orbit: Washington, D.C., U.S. Government Printing Office, NASA SP-362, 255 p.

Maxwell, T. A., El-Baz, F., and Ward, S. H., 1975, Distribution, morphology, and origin of ridges and arches in Mare Serenitatis: Geological Society of America Bulletin, v. 68, p. 1273-1278.

Menzel, D. H., and others, 1971, Report on lunar nomenclature: Space Science Reviews, v. 12, p. 136–186.

Mutch, T. A., 1972, Geology of the Moon—A stratigraphic view: Princeton, New Jersey, Princeton University Press, 391 p.

Pike, R. J., 1974, Depth/diameter relations of fresh lunar craters: Geophysical Research Letters, v. 1, p. 291-294.

——1975, Size-morphology relations of lunar craters: Discussion: Modern Geology, v. 5, p. 169–173.

Shoemaker, E. M., 1960, Penetration mechanics of high velocity meteorites, illustrated by Meteor Crater, Arizona: International Geological Congress, 21st, Copenhagen, v. 18, p. 418–434.

——1962, Interpretation of lunar craters, in Kopal, Z., ed., Physics and astronomy of the Moon: London, Academic Press, p. 283–359.

Shoemaker, E. M., and Hackman, R. J., 1962, Stratigraphic basis for a lunar time scale, in Kopal, Z., and Mikhailov, eds., The Moon: London, Academic Press, p. 289–300.

Stuart-Alexander, D. C., and Wilhelms, D. E., 1974, Nectarian System, a new lunar stratigraphic name: U.S. Geological Survey Journal of Research, v. 3, p. 53–58.

Wilhelms, D. E., 1970, Summary of lunar stratigraphy—Telescopic observations: U.S. Geological Survey Professional Paper 599-F, 47 p.

Wilhelms, D. E., and El-Baz, F., 1977, Geologic map of the east side of the Moon: U.S. Geological Survey, Miscellaneous Investigation Series, Map I-948.

Wilhelms, D. E., and McCauley, J. F., 1971, Geologic map of the near side of the Moon: U.S. Geological Survey, Miscellaneous Investigation Series, Map I-703.

MANUSCRIPT RECEIVED BY THE SOCIETY SEPTEMBER 5, 1979
MANUSCRIPT ACCEPTED MAY 20, 1980

Geological Society of America
Special Paper 183
1980

G. K. Gilbert and ground water, or
'I have drawn this map with much reluctance'

ALLEN F. AGNEW
Senior Specialist in Environmental Policy (Mining and Mineral Resources), Congressional Research Service, Library of Congress, Washington, D.C.

ABSTRACT

Gilbert's studies of underground water were overshadowed by his magnificent reports on geologic structure and landforms. In fact, his 1896 publication on the underground water of the Arkansas Valley in eastern Colorado was called by W. M. Davis "for the most part a straightforward geological account of the successive strata."

Gilbert's 1896 report devoted equal space to stratigraphy and to underground water. His discussion of artesian water was based largely on Chamberlin's excellent paper on that subject, a decade previously; he briefly acknowledged his debt to Chamberlin for the section dealing with the general occurrence of artesian water, and acknowledged more completely his debt to his U.S. Geological Survey colleague, F. H. Newell.

Gilbert's reports of 1896 and 1897 are largely practical ones on where and how to explore for artesian water, and they include maps showing areas where water can be expected in wells at different depths. Plagued by too little information, he was courageous enough to put lines on a map, although he admitted that the data "are too imperfect to fix the lines definitely except at a few points." Thus, he said, "I confess that I have drawn this map with much reluctance." He recommended experimental borings, to gather the additional information needed before putting down a well.

Gilbert intended his text and maps to be read by local residents (nongeologists) in their search for artesian water supplies, and this is seen as a major contribution of his work on the underground water of the Arkansas Valley.

A TEXTBOOK

Dakota Formation. . .its sandstones constitute the more important reservoir of artesian water in Western North America. [Gilbert, 1893]

This quotation, published in *Johnson's Universal Cyclopedia,* may not have been the first and certainly was not the last comment on ground water by G. K. Gilbert. However, it was one of relatively few such utterances by him, for Gilbert was far more interested in surface water and its effects in causing and modifying landforms.

A decade after publication of the *Cyclopedia,* Gilbert wrote briefly about ground water and its effects in a textbook on physical geography. Discussing underground changes in the Earth's crust, Gilbert and Brigham (1904, p. 96) noted that "ground water, coming in contact with the minerals that make up the rocks, is able to take more or less of the substances into solution." This process leads to the "destruction of rocks by underground waters" (Gilbert and Brigham, 1904, p. 98).

This textbook also discussed, albeit briefly, wells, artesian wells, and water supply. The authors recognized that the rock and unconsolidated material near the surface, which they termed "the mantle of waste," were "usually filled with water except the uppermost part." This ground water, as they called it, "supplies ordinary springs and is itself replenished by the part of the rain which soaks in." A well, they commented, is a boring or digging that "reaches below the level of permanent ground-water;" water stands in the well up to the ground-water level, and as it is pumped out, more water is "restored by oozing from the sides" (Gilbert and Brigham, 1904, p. 103).

They wrote about artesian wells, noting that these are named from the province of Artois in France. They presented a cross section showing the artesian plumbing system as first described by T. C. Chamberlin (1885), although, strangely, they did not credit him. Gilbert and Brigham (1904, p. 104) remarked that this general arrangement of interlayered "porous and compact beds is fortunately found in many regions. Hence along the Atlantic Coast, as in southern New Jersey, such wells are common, also in northern Illinois about Chicago, and in many parts of the Great Plains from the Dakotas to Texas."

These textbook writers understood how easy it is to pollute one's water supply, whether from springs, wells, rivers, or lakes. They commented: "Unfortunately, not all water is wholesome, and much that is used is dangerous to health and life. As the rain soaks down to join the ground-water it may carry with it any filth that lies on or near the surface, and thus make springs and wells unfit to use" (Gilbert and Brigham, 1904, p. 104). They cautioned that "everyone should learn about the movements of ground water to arouse caution in the use of wells, and everyone should understand that water which is perfectly transparent and pleasant to the taste may at the same time be filled with the germs of typhoid fever and other diseases" (Gilbert and Brigham, 1904, p. 105). Furthermore, they applauded the fact that increasing attention was given to the important subject of pollution of water supplies, by cities and towns and by the State and National Governments.

THE BRIDGE

The classic biography of G. K. Gilbert by Davis (1927) recounted Gilbert's work in the Arkansas Valley. Davis's view of geology must have influenced him to characterize Gilbert's report as "for the most part a straightforward geological account of the successive strata in the district treated" and as an "altogether new conception of the physiography of the Rocky Mountain region during Tertiary time." Davis continued:

Much the most important product of Gilbert's work in Colorado was the recognition of a largely fluviatile origin for the fresh-water Tertiary strata of the plains, which, like similar strata in basins among the mountains farther west, had been universally treated as of lacustrine origin by all other geologists. . . . This was the beginning of an altogether new conception of the physiography of the Rocky Mountain region during Tertiary time. [Davis, 1927, p. 207]

Davis went on to note items discussed in writings of Gilbert, based on the latter's studies of the Plains, as including: (1) measurement of Cretaceous time, (2) some small laccoliths, (3) Tepee buttes, (4) fire clays, and (5) average composition of sedimentary rocks. Davis lamented the fact that "all these subordinate problems are good enough in their way, but they were rather random and discontinuous efforts" (Davis, 1927, p. 208). Gilbert's sections on artesian water, or water in the Dakota Formation, and on ground water in his report on the Arkansas Valley of eastern Colorado (Gilbert, 1896, p. 30–51), which actually gave substance to the title of the whole report, apparently were seen by Davis as being even more "subordinate," for he referred to them not at all.

However, when commenting on Gilbert's Pueblo geologic folio, published a year later (Gilbert, 1897), Davis did give passing recognition to Gilbert's ground-water work:

Besides the standard series of maps the Pueblo folio contains a stereogram. .. representing the warped and faulted surface of a standard stratum, uncovered where still buried and restored where already eroded, and *also a special map indicating by underground contours the depth of the uppermost water-bearing strata* [emphasis added]. [Davis, 1927, p. 208]

Thus, whereas Gilbert's structural geology caught Davis's eye, that same eye was virtually impervious to Gilbert's work on ground water. It is the latter, which Davis must have regarded as even less significant than "subordinate" among Gilbert's "random and discontinuous efforts," to which the remainder of this account is devoted.

THE ARKANSAS VALLEY REPORT

Gilbert gave a straightforward account of the geology in his 1896 paper, "Underground Water of the Arkansas Valley in Eastern Colorado." He devoted 20 pages to the stratigraphy of the consolidated and unconsolidated materials of the area, with brief comment on both the topography and the geologic structure. He next discussed artesian water for 15 pages, and spent the final 5 pages considering ground water.

Gilbert was careful to acknowledge his debt to his colleague, F. H. Newell, noting that he had conferred frequently with Newell during the preparation of the manuscript and that he had "drawn freely on his [Newell's] expert knowledge of the general subject of water supply" (Gilbert, 1896, p. 51).

Frederick Haynes Newell, having joined the Survey in 1888 as assistant hydraulic engineer, was the first full-time appointee on Powell's Irrigation Survey; he headed the Reclamation Service of the Survey in 1902 and, when it became a separate agency (today known as the U.S. Bureau of Reclamation), served as its director during the period 1907 to 1914. He was chief hydrographer of the Survey during the period 1895 to 1902. Although Newell did not publish water-supply reports himself, he guided and nurtured numerous Survey geologists while they carried out their investigations of such topics. Thus, Gilbert had selected a worthwhile colleague for his study of the underground water of the Arkansas Valley in eastern Colorado—and by this acknowledgment he expressed his appreciation to his colleague, Newell.

Artesian Water

Much of Gilbert's discussion of artesian water appears to have been derived from the excellent work of Chamberlin (1885). (Actually, Chamberlin had described the principles of artesian water earlier, but all who have cited his fundamental work on this subject have neglected to mention his 1883 publication.) Gilbert (1896, p. 32) acknowledged that "the natural circumstances affecting the availability of artesian water are ably and comprehensively treated" by Chamberlin.

Gilbert (1896, p. 30) described two classes of underground water, which today are termed "unconfined" and "confined." Unconfined water, he said, "flow[s] through a porous bed which is continuous to the surface of the ground, in which case the position of its upper surface [today called the "water table"] varies with the supply." Regarding confined water, Gilbert (1896, p. 31) showed that water which "flow[s] through a porous stratum confined between impervious strata. . .usually occupies the entire stratum and presses [both downward and upward] against the impervious [beds]." When water of the unconfined type "is reached by a well it retains its natural level within the well." When a well is put down to confined water, however, "the water rises somewhat within the opening, the height. . .determined by its original pressure against the covering rock."

Gilbert used terminology which has been modified for use today; he commented that unconfined water of the first type is called "ground-water," and confined water, which today we include as part of ground water, he called "artesian." He took note of the disagreement by some people with his use of "artesian," saying that they would restrict the term to waters that flow at the surface. Gilbert held, as we do today, that artesian water can be divided into "flowing," when the water is naturally discharged at the surface of the ground, and "pumping," when it rises in the well bore but not to the surface.[1]

----

[1]A year later in the Pueblo geologic folio (Gilbert, 1897, p. [7]), he used the term "head," and defined it as the highest level to which water in an artesian well will rise. "If the head is above the surface of the ground the well is said to flow; otherwise it is called a pumping well," he said.

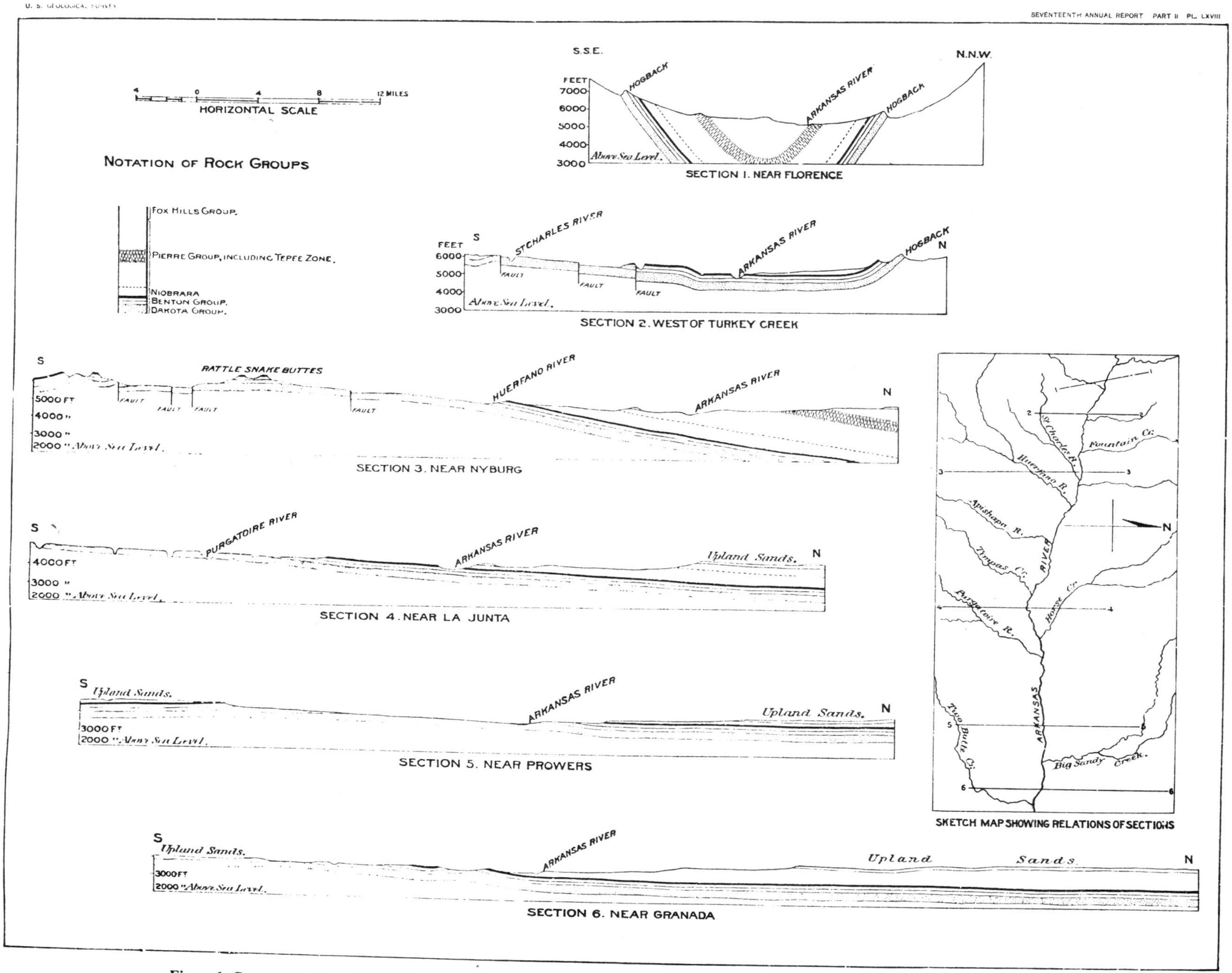

Figure 1. General structure of the Dakota Formation and associated rocks in the Arkansas Valley district (Gilbert, 1896, Pl. LXVIII).

Gilbert (1896, p. 31) went on to discuss the *general conditions* that must be fulfilled for artesian water to occur—the two elements that Chamberlin (1885) had discussed so thoroughly and diagrammed a decade earlier. These elements were that the porous bed, called "aquifer" today, must "receive its supply at a point or in a region where it lies comparatively high," and it must "be enclosed by comparatively impervious beds."

Gilbert had an eye for the economics of such water supply, because he remarked:

The practical value of a body of artesian water depends on the quantity which can be continuously supplied to wells, and this, in turn, on the size and character of the conduit, on the freedom with which water is received in the region of imbibition, and in some cases on the rainfall. [Gilbert, 1896, p. 31]

Besides noting the importance of the good quality of artesian water, Gilbert (1896, p. 31–32) considered the value of artesian water as being dependent on its depth beneath the surface and "on other factors affecting the cost of securing it."

In the Arkansas Valley of Colorado, Gilbert observed that "the only artesian water of demonstrated value is that contained in the Dakota sandstone." He described the *gathering grounds* as being in the hogbacks to the west, and illustrated their general structural configuration in a series of cross sections. (Fig. 1).

Gilbert described the underground plumbing system with considerable acumen, and noted that the fine texture of the upper layers of the Dakota Formation inhibits its receiving and transmitting much water. As a consequence, he was unable to make specific estimates of the amount of precipitation that "may annually be imbibed," either directly from rainfall or indirectly by infiltration downward from the overlying upland sands (Gilbert, 1896 p. 33). Regarding *capacity*, however, he commented that the "capacity of the Dakota [or of any other artesian bed] depends on the thickness of the water-bearing sandstone strata and on their porosity." Further, he said,

The amount of water flowing from a well depends in part on the height of the point of discharge as compared to the head of the water. . . , but the amount which may be obtained by pumping depends entirely on the capacity of the rock as a conduit—that is, on its thickness and the resistance which its texture opposes to the free flow of water. [Gilbert, 1896, p. 34]

The problem of overdraft—that is, the mining of ground water because of the excessive number of wells in use—was described forthrightly: "Every well from which artesian water is drawn diminishes the head or pressure in the neighboring portions of the water-bearing formations so that a well bored near it can not obtain so much water as it otherwise might" (Gilbert, 1896, p. 35). As the artesian pressure decreases, the flowing wells no longer flow and must be pumped, and the cost of such pumping gradually increases. Thus, Gilbert (1896, p. 35) warned that "if the inhabitants of this district are wise they will profit by the experience of other communities and limit their expectation as to the quantity of water which can be derived from the artesian source beneath."

Regarding *distribution*, or occurrence, of artesian water, Gilbert prepared a map of the artesian district (Fig. 2) showing by two patterns the areas where "the depth of water is estimated as less than 1000 feet, and. . .[where it] is estimated as between 1000 and 2000 feet." He admitted that "the data at hand are too imperfect to fix the lines definitely except at a few points"; therefore, "because it is impossible in such delineation to give adequate expression to doubt, . . . I confess that *I have drawn this map with much reluctance* [emphasis added]" (Gilbert, 1896, p. 36).

Gilbert's becoming modesty did not blind the residents of the Arkansas Valley district to the value of his work. As reported by the director of the Survey in his Annual Report (Walcott, 1896, p. 74–75): "The results obtained by [Gilbert from his field work] have had immediate practical benefit, as shown by the confidence with which deep drilling has been undertaken at various points."

One would hope that others who have faced this same problem of inadequate data have also had the courage to go ahead and present their best interpretations of the facts at hand, based upon their backgrounds and experience. That willingness to commit oneself where there is yet some degree of doubt, together with the humility to acknowledge this doubt, is a mark of true professionalism.

Gilbert explained his doubts and also incorporated a warning to the reader, as follows:

When the region has been thoroughly surveyed, the free-curving boundary lines of each belt as here drawn will be replaced by more sinuous lines, and doubtless the general position of each one will be in some parts materially changed. The reader is warned that he must place no reliance on the local details of the map, but regard it merely as the expression of certain general facts as to the accessibility of the artesian water. [Gilbert, 1896, p. 36]

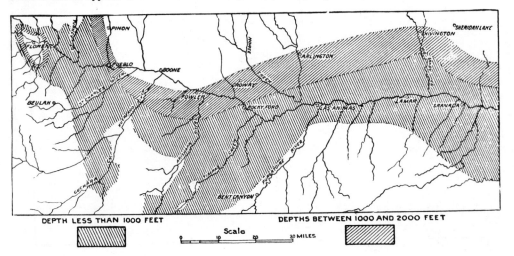

DEPTH LESS THAN 1000 FEET   DEPTHS BETWEEN 1000 AND 2000 FEET

Scale

0   10   20   30 MILES

Figure 2. Map of the Arkansas Valley artesian district showing the areas in which water occurs in the Dakota Formation at different depths (Gilbert, 1896, Fig. 49).

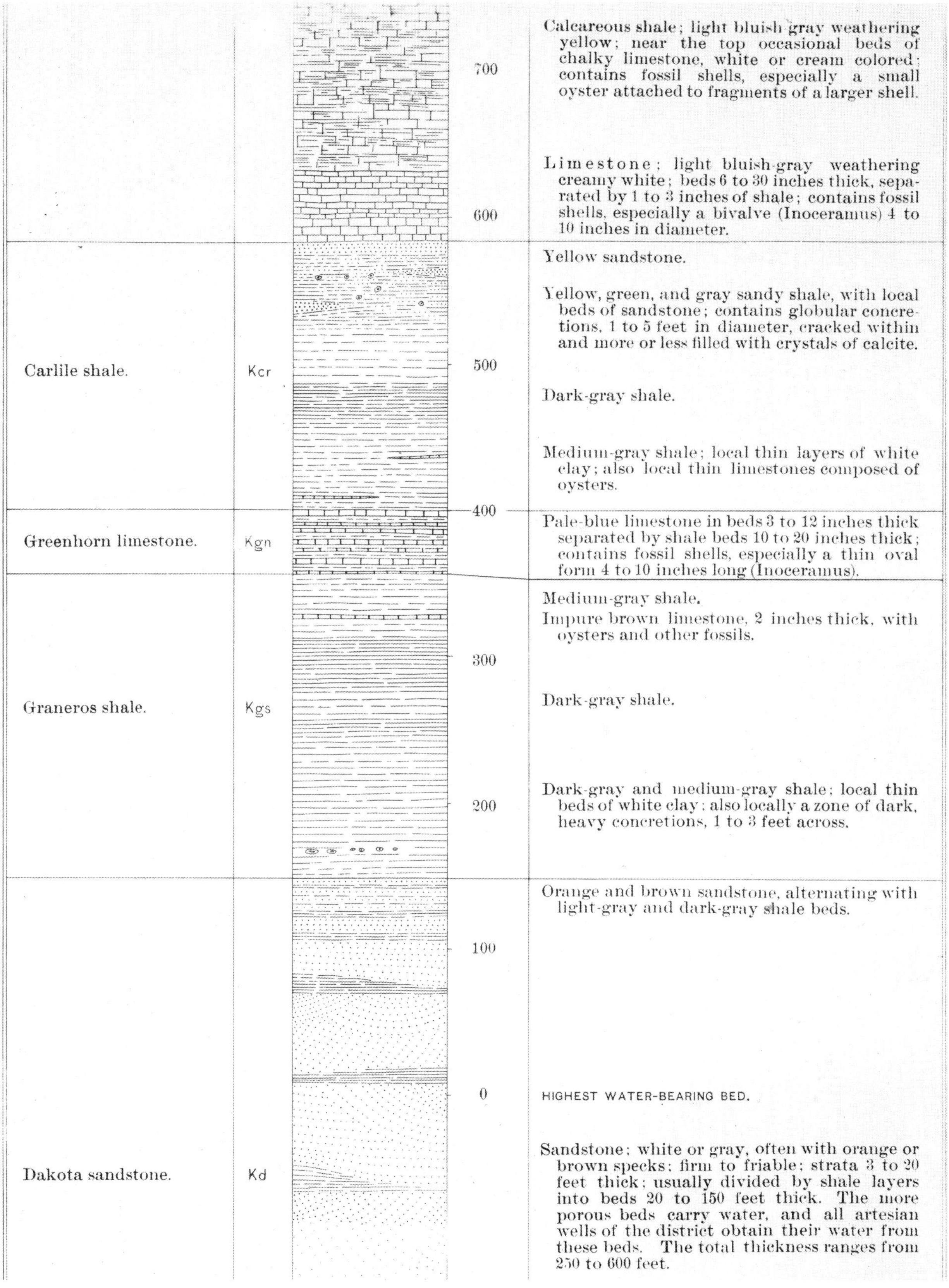

				Calcareous shale; light bluish-gray weathering yellow; near the top occasional beds of chalky limestone, white or cream colored; contains fossil shells, especially a small oyster attached to fragments of a larger shell.

700

Limestone; light bluish-gray weathering creamy white; beds 6 to 30 inches thick, separated by 1 to 3 inches of shale; contains fossil shells, especially a bivalve (Inoceramus) 4 to 10 inches in diameter.

600

Carlile shale.   Kcr

Yellow sandstone.

Yellow, green, and gray sandy shale, with local beds of sandstone; contains globular concretions, 1 to 5 feet in diameter, cracked within and more or less filled with crystals of calcite.

500

Dark-gray shale.

Medium-gray shale; local thin layers of white clay; also local thin limestones composed of oysters.

400

Greenhorn limestone.   Kgn

Pale-blue limestone in beds 3 to 12 inches thick separated by shale beds 10 to 20 inches thick; contains fossil shells, especially a thin oval form 4 to 10 inches long (Inoceramus).

Medium-gray shale.
Impure brown limestone, 2 inches thick, with oysters and other fossils.

300

Graneros shale.   Kgs

Dark-gray shale.

Dark-gray and medium-gray shale; local thin beds of white clay; also locally a zone of dark, heavy concretions, 1 to 3 feet across.

200

Orange and brown sandstone, alternating with light-gray and dark-gray shale beds.

100

0   HIGHEST WATER-BEARING BED.

Dakota sandstone.   Kd

Sandstone; white or gray, often with orange or brown specks; firm to friable; strata 3 to 20 feet thick; usually divided by shale layers into beds 20 to 150 feet thick. The more porous beds carry water, and all artesian wells of the district obtain their water from these beds. The total thickness ranges from 250 to 600 feet.

Figure 3. Part of the stratigraphic column in the Arkansas Valley artesian district showing the thickness of rocks above the Dakota Formation (Gilbert, 1897, artesian column).

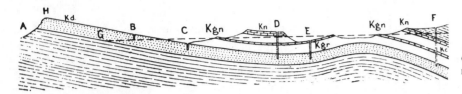

Figure 4. Idealized cross section showing the different artesian conditions in the Pueblo Quadrangle (Gilbert, 1897, Fig. 16).

With regard to potentially flowing artesian wells, Gilbert remarked that determining such areas is a more difficult matter because "it depends partly on the comparative porosity of the rock..., a character which can not be directly observed." For this reason, he recommended "experimental boring"—just as we do today.

However, in regard to the *quality* of artesian water, Gilbert was a victim of his time. He stated that a "common character of artesian waters" is that they are "probably entirely free from living organisms..., which are effectually filtered in their long passage through the rocks" (Gilbert, 1896, p. 37). This generalization followed his statement that "the water derived from the Dakota sandstone is practically free from organic matter." Gilbert was aware by 1904, some 7 yr later, that the quality of water from the nonartesian beds could be "dangerous to health and life [as] it may carry any filth that lies on or near the surface...as it soaks down to join the ground-water."

Gilbert was aware of the variations in chemical quality of waters in the Dakota Sandstone, for he remarked that "the mineral impurity of the Dakota water is...5 to 15 times as great as in the...great rivers of the country." Four analyses, which showed 888 to 2500 ppm of "dissolved minerals," placed the Dakota water "in the class of mineral waters," he said. Both the character and quantity of mineral impurities varied from place to place, and also from bed to bed as the different sandstone layers were penetrated.

Two tables of water analyses for the four wells were presented by Gilbert, one as ions and other as salts. Gilbert (1896, p. 40–44) commented that these analyses came from various sources and were originally reported in various ways. "To make them comparable they have been reduced to uniformity in mode of statement, and in part recomputed. . . . The second [salts] is the more usual form of statement, but the first [ions] is of higher authority." He acknowledged that in the arrangement of the tables and in the discussion of their data he was especially indebted to F. W. Clarke and W. G. Hillebrand of the chemical division of the Survey for their "expert knowledge and friendly aid."

Gilbert (1896, p.42) related the chemical quality of the Dakota water to potential water uses, by commenting that "the total mineral content is so large as to produce an appreciable and ultimately deleterious deposit in irrigated soil, unless occasionally leached out or neutralized by corrective action," and observing "the comparative tendency of the water to form deposits in boilers."

Next, Gilbert (1896, p. 42–43) turned to *prediction*. He remarked that the cost of making—that is, drilling and completing—an artesian well depended largely on the thickness of rock that must be penetrated in order to reach the water, together with the depth and whether the well would flow. "Predictions as to these [latter] two points will doubtless have greatest value if made by a geologist who has personally studied the district," he said, thus indicating the value of a trained and capable professional.

However, Gilbert (1896, p. 43) ensured that his report would possess even greater value to the residents of the 1,500 mi²-area covered by his map by assuring them that "in many cases the resident can be his own geologist, and it is worthwhile to point out how he can help himself to the knowledge he desires." He thereupon proceeded for the next two and a half pages to lead the nongeologist through a review of the general arrangement of the rocks as shown in the cross sections in his Plate LXVIII (Fig. 1 herein); some general facts about the depth of artesian water as shown in his Figure 49 (Fig. 2 herein); and, with the aid of his Figure 46, estimating how deep one would have to drill "in order to reach the first water-bearing sandstone."

This is an excellent example, from 80 years ago, of a geologist communicating with the nongeologic public. Through the years there have been many other examples of such outreach from the professional scientist to the nonscientist by members of the Survey as well as the numerous State Geological Surveys. Today, when nongeologists need more understanding of geologic principles and processes and should be more aware of the value of this knowledge in solving many societal problems, one is all the more appreciative of Gilbert's wisdom and his desire to communicate with those "other" publics.

Ground Water

The term "ground water" was restricted by Gilbert to the nonartesian part of underground water, as was common usage before the benchmark publications of Meinzer (1923a, 1923b). Although Gilbert devoted twice as many printed pages to artesian water as to ground water, he commented that artesian flow is the exception rather than the rule, whereas "the conditions favorable to the accumulation of ground water are widespread." Much of the rain that falls on the ground—that which escapes evaporation or immediate runoff at the surface—"finds its way gradually downward until it meets with some impervious formation, [whereupon it] either remains stagnant as a sort of subterranean lake, or flows slowly in some direction, as a sort of subterranean stream" (Gilbert, 1896, p. 45).

Today, Gilbert's use of "subterranean stream" is viewed as unfortunate, although it was then the common expression of geologists and hydrologists. Unfortunate, because the legal profession has clung to that term (with all its sixteenth century mystery and misconceptions) as the major anchor in debating the law of water rights and water use as applied to underground waters. Gilbert and his colleagues conceived of the movement of the water through unconsolidated sediments as being dependent on permeability and thus varying in quantity and rate according to local conditions in the sediments, although they used the term "porosity" in this sense. Unfortunately, the larger public of 80 years ago (and even today) envisioned underground streams rushing along as they do on the surface of the ground. Today we know that such underground flow occurs only in open conduits like those in cavernous limestones or gypsum beds; except in localized areas where such strata constitute the bedrock, such underground flow is a rarity.

Largely through the efforts of Federal and State ground-water geologists and hydrologists in the past decade, many states have been able to revise their water laws to incorporate a more accurate concept and statement of the hydrology of water underground in contrast to water on the surface. Thus, the century-old water-rights/water-law confrontation between proponents of "diffuse flow" and "underground streams" is gradually being alleviated by modern scientific and engineering knowledge.

Gilbert commented that "the quantity of water in the ground, its depth beneath the surface, its quality and its direction and rate of its flow depend on a variety of circumstances," which included rainfall and the porosity of the soil. He used the term "water table" properly as being the upper limit of the ground water, noting that it marks the height at which water will stand in a well; it is the surface separating saturated earth from earth that is merely moist (Gilbert, 1896, p. 46).

Gilbert's knowledge of ground-water hydrology was very thorough, for he stated that the water table is not a level surface, but slopes in the direction of ground-water flow. Indeed, he pointed out that "water plain" was incorrectly used for water table. The flow rate of water underground is "very slow as compared to the flow of a surface stream, [and depends] on the great friction produced by the movement of water through narrow passages." He noted that the quality of ground water also varies from place to place.

Gilbert described the *upland sands* as constituting the most important reservoir of ground water in the Arkansas Valley district.[2] In discussing the variable quality of such ground water, he correctly ascribed the differences as depending somewhat on the character of the underlying rock, but erroneously stated that the local differences in quality are "probably more closely related to the volume of flow." This may have been a slip of the pen whereby he had meant to say "rate" rather than "volume," for he continued that "where the flow is comparatively rapid the chlorides and more soluble sulphates have already been leached out; but where the currents are feeble the leaching is still in progress" (Gilbert, 1896, p. 47). His term "current" was unfortunate, for it again conjures up visions of flowing underground streams.

Other sources of ground water that Gilbert described in the area were the sands of the Arkansas River Valley *terraces* and the windblown *dune sands*, both of which received precipitation, and *underflow* from the larger streams "through bottom lands of loam, sand, and gravel" (Gilbert, 1896, p. 48). He noted that the quality of water from the terraces and the underflow sediments is variable. The terrace water is modified by material leached from cappings of sand and gravel as well as from underlying shales; on the other hand, the water of the longer streams is "purer near its source than after it has flowed for a greater distance and been reduced by evaporation."

Gilbert commented that irrigation causes deterioration "because of the mineral matter brought from the irrigated soils and subsoils." He was prescient in this, as is borne out by the decline in productivity of such soils by irrigation practices.

---

[2] In the Pueblo geologic folio he said that "the only important water-bearing beds in the district are the sandstones of the Dakota formation" (Gilbert, 1897, p. [7]). His term "beds" apparently did not include the upland sands; also, the Dakota water is artesian, whereas the ground water in the upland sands is not.

THE PUEBLO FOLIO—A "HOW-TO" HANDBOOK

A year later, Gilbert's scientific study of part of the Arkansas Valley was also published as the Pueblo geologic folio (Gilbert, 1897). He made this a handbook, to be used by local searchers for ground water, for he said (p. [7]), "It is assumed. . .that the facts and deductions recorded in the folio [are] of interest not only to professional geologists but to those residents of Pueblo and vicinity who care to know the origin of rocks and hills or who desire to make use of the mineral products of the district." In noting that the Dakota Formation is absent from certain small areas, he explained that it occurs throughout most of the quadrangle but is buried under other rocks, chiefly shales. Accordingly, "owing to the deformation of the rocks and the resulting unequal erosion, its depth below the surface [ranges] from nothing to more than 3000 feet."

Gilbert presented a stratigraphic column (Fig. 3) so that those who prospected for ground water would know the kinds and thicknesses of rock units that they would penetrate before reaching the Dakota Formation, and those who started a well where the surface rock was the top of the Greenhorn Limestone would know that they must penetrate all 50 ft of that formation and 200 ft of the underlying Graneros Shale, in addition to the 100 ft or more of Dakota beds. The latter point shows that Gilbert knew that the Dakota Formation was not uniformly water-bearing throughout. He noted that the water-bearing beds of the Dakota "occur at different distances from the top [of the formation] in different localities, the depths of the first or highest being in places less than 100 feet, and elsewhere more than 150 feet." The number of such water-bearing beds, Gilbert said, "probably ranges from one to three or four, but this has not been tested by the drill, as the borer usually stops at the first good stream" (Here again is an example of Gilbert's unfortunate mention of the underground stream of water.) In his columnar sections illustrating this stratal sequence, Gilbert placed the zero for the scale in feet "at the horizon of the highest water-bearing bed. [Thus,] if one can determine what member of the rock system forms the surface of the ground at any spot, one can read in this section the depth to water." This should have been an excellent aid to the ground-water seeker. In addition to the columnar section, Gilbert presented an idealized cross section which showed these relations for well bores in various geologic structural situations (Fig. 4).

Gilbert then explained the contour map that he had drawn[3] (Figs. 5, 6). "Each contour is a line drawn through points where the estimated depth of water below the surface of the ground is the same." Noting that the formations above the Dakota "are remarkably uniform in thickness from place to place, so that a thorough knowledge of the rock formations makes it possible to predict with tolerable accuracy the depth at which the first Dakota stratum will be encountered," Gilbert went on to say that "the Dakota formation itself is much more variable, not only in thickness but in the order and number of its water-bearing beds."

Gilbert's confidence in this contour map is shown by his statement that "while the estimates [of depth to water in the Dakota] are everywhere subject to some error, especially from the variability of the Dakota formation, it is believed that they will in general come within 100 to 200 feet of the fact, and thus prove practically serviceable." He indicated the areas of greatest uncer-

---

[3] It was this map that had caught the eye of Gilbert's biographer (Davis, 1927, p. 208).

Figure 5 (facing page). Map of the northeastern corner of the Pueblo Quadrangle showing by contours the depth to water, and areas of water yield according to four types of availability (Gilbert, 1897, artesian water sheet).

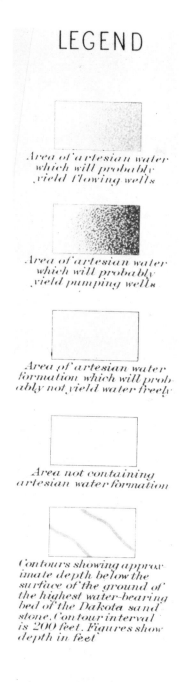

LEGEND

Area of artesian water which will probably yield flowing wells

Area of artesian water which will probably yield pumping wells

Area of artesian water formation which will probably not yield water freely

Area not containing artesian water formation

Contours showing approximate depth below the surface of the ground of the highest water-bearing bed of the Dakota sandstone. Contour interval is 200 feet. Figures show depth in feet

✗ Artesian wells

tainty, caused by the lack of good exposures, by drawing the contours as broken instead of solid lines.

Next, Gilbert turned to a discussion of "head" (today, we define this term as including the effects of altitude, velocity, and pressure), as it determines whether a well will flow or will have to be pumped. Reminding his readers that the water in the rocks does not lie motionless (and thus would have the same head everywhere),

Gilbert stated again that the water is "really flowing at a slow rate through the rock, and its head varies from point to point."

Gilbert considered prediction of head to be uncertain, because there was little information on the various factors contributing to the mode and rate of variation of flow, the source of supply, the direction of flow, and the resistance to flow through small pores. Furthermore, he observed perceptively that the head is usually not the same for the different water-bearing beds. Still more importantly, the head is reduced in each locality by every draft on the supply through a flowing or pumping well. For these reasons, "the line on the map separating the supposed flowing from the supposed pumping territory *was drawn with much less confidence than the contours of depth* [emphasis added]" (Gilbert, 1897, p. [7]).

GILBERT, THE PIONEER

Gilbert's reports of 1896 and 1897, which dealt with ground water (including artesian water) appear to have been the first to apply the knowledge imparted by Chamberlin's excellent paper of a decade earlier. Gilbert's method of portraying ground-water availability by the use of maps was adopted by others in the Survey, as shown by the Folios by Hills (1899, 1900, 1901) in Colorado, the Water-Supply Paper by Haworth (1897) in Kansas, and the Water-Supply Papers by Russell (1897, 1901, 1903) in Washington, Idaho, and Oregon. Haworth acknowledged his debt to Gilbert (and also to Darton, 1896), and Russell in 1897 credited Chamberlin and in 1901 cited both Chamberlin and Gilbert. Gilbert was also praised by Newell (1899) for his "very thorough description" of the large artesian basin of the plains region.

On the other hand, Hills did not acknowledge this pathfinding work of Chamberlin and Gilbert, although he prepared artesian water sheets according to Gilbert and cross sections according to Chamberlin. Others in the Survey at that time also seem to have profited by the work of these two forward thinkers, but without acknowledging indebtedness to them; for example, Lindgren (1898), whose Boise folio showed good understanding of artesian conditions, and Darton (1896, 1897, 1901, 1902, 1903, 1904a, 1904b, 1905, 1906), whose Annual Reports, Folios, and Professional Papers went beyond those of Gilbert in discussing ground water. In fact, it is most surprising that Darton's (1906) extension of Gilbert's (1896) work on the Arkansas Valley for 250 mi to the east did not refer to either Gilbert or Chamberlin. (Gilbert had been in charge of Darton's work earlier in Darton's career, in the eastern part of the United States. Could this have contributed to that nonrecognition?)

Thus, Gilbert's pioneering work anticipated that of others in several series of Survey publications on water: Water-Supply Papers, Professional Papers, and Folios.

CONCLUSIONS

It is evident that Gilbert was not reluctant about *some* of the lines on his maps, particularly the contours showing depth to water; he was, understandably, less confident about the line separating the areas where pumping wells could be drilled relative to nonpumping (or flowing) ones.

It is obvious, in both the Pueblo Folio and the Arkansas Valley report, that he had absorbed well the philosophy of artesian flow that was pioneered by T. C. Chamberlin, and that he had profited

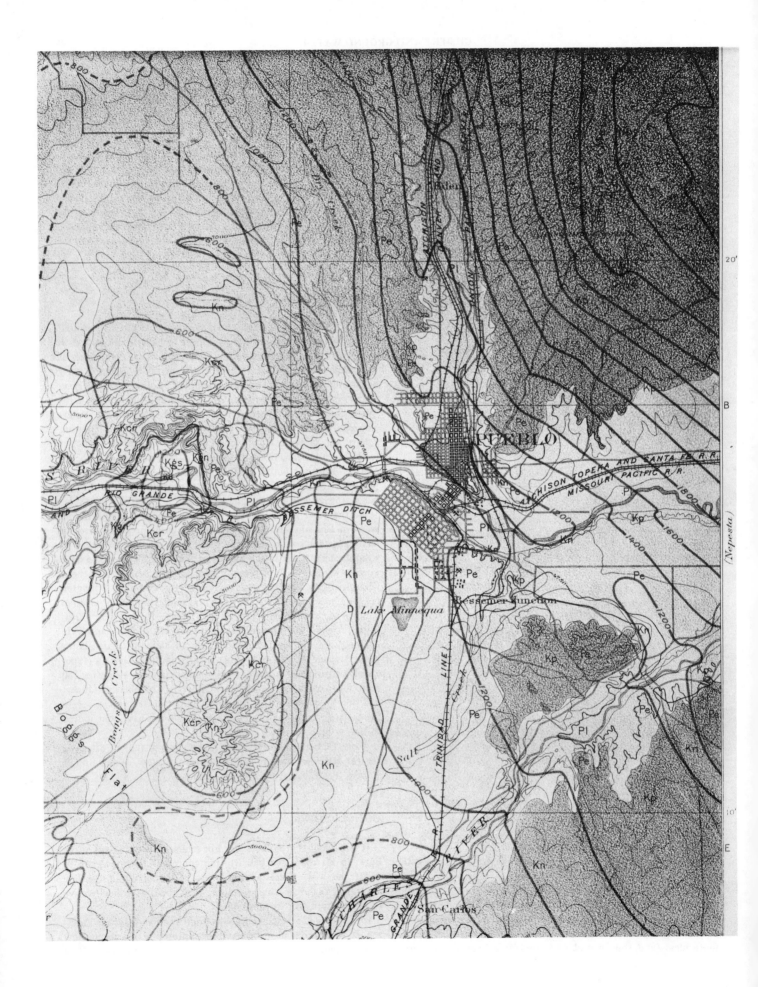

Figure 6 (facing page). Map of the central part of the Pueblo Quadrangle showing the areas of closely spaced and widely spaced contours (Gilbert, 1897, artesian water sheet).

by his association with F. H. Newell. In knowing whom to follow in the matter of ground-water hydrology, G. K. Gilbert was a most perceptive scientist.

Like many of his contemporaries of the latter part of the nineteenth century, Gilbert was a masterful communicator, both in word and in illustration. But it seems to me that his unique contribution to the study of ground water was his willingness and ability to communicate with nongeologists, so that such people could reap a greater benefit from the scientific work that Gilbert and his associates had performed for the Survey.

It is true that G. K. Gilbert may have drawn *one* line on his map with great reluctance. Nonetheless, in both reports he had qualified it, and he had shown why and how he had drawn his lines, as well as the geology on which all of his other conclusions were based.

We need have no reluctance at all in recognizing the ground-water work of this giant, G. K. Gilbert, as being both pioneering and truly professional, and in the best tradition of the U.S. Geological Survey.

REFERENCES CITED

Chamberlin, T. C., 1883, Artesian wells, *in* Geology of Wisconsin, v. I, p. 689–701.

——1885, The requisite and qualifying conditions of artesian wells: U.S. Geological Survey 5th Annual Report, p. 125–173.

Darton, N. H., 1896, Preliminary report on the artesian waters of part of the Dakotas: U.S. Geological Survey 17th Annual Report, pt. II, p. 603–694.

——1897, New developments in well boring and irrigation in eastern South Dakota: U.S. Geological Survey 18th Annual Report, pt. IV, p. 561–615.

——1901, Preliminary description of the geology and water resources of the southern half of the Black Hills and adjoining regions in South Dakota and Wyoming: U.S. Geological Survey 21st Annual Report, pt. IV, p. 489–599.

——1902, Oelrichs folio, South Dakota-Nebraska: U.S. Geological Survey Geologic Atlas, Folio 85, 6 p.

——1903, Preliminary report on the geology and water resources of Nebraska west of the 103d meridian: U.S. Geological Survey Professional Paper 17, 69 p.

——1904a, Newcastle folio, Wyoming-South Dakota: U.S. Geological Survey Geologic Atlas, Folio 107, 9 p.

——1904b, Edgemont folio, South Dakota-Nebraska: U.S. Geological

Survey Geologic Atlas, Folio 108, 10 p.

——1905, Preliminary report on the geology and underground water resources of the central Great Plains: U.S. Geological Survey Professional Paper 32, 433 p.

——1906, Geology and underground waters of the Arkansas Valley in eastern Colorado: U.S. Geological Survey Professional Paper 52, 90 p.

Davis, William M., 1927, Biographical memoir of Grove Karl Gilbert, 1843–1918: National Academy of Sciences, v. XXI (5th Memoir), 300 p.

Gilbert, Grove K., 1893, Dakota Formation: Johnson's Universal Cyclopedia, v. II, p. 639.

——1896, Underground water of the Arkansas Valley in eastern Colorado: U.S. Geological Survey 17th Annual Report, pt. II, p. 557–601.

——1897, Pueblo folio, Colorado: U.S. Geological Survey Geologic Atlas, Folio 36, 7 p.

Gilbert, Grove K., and Brigham, Albert P., 1904, An introduction to physical geography: New York, D. Appleton & Company, 380 p.

Haworth, Erasmus, 1897, Underground waters of south-western Kansas: U.S. Geological Survey Water-Supply Paper 6, 65 p.

Hills, R. C., 1899, Elmoro folio, Colorado: U.S. Geological Survey Geologic Atlas, Folio 58, 6 p.

——1900, Walsenburg folio, Colorado: U.S. Geological Survey Geologic Atlas, Folio 68, 6 p.

——1901, Spanish Peaks folio, Colorado: U.S. Geological Survey Geologic Atlas, Folio 71, 7 p.

Lindgren, Waldemar, 1898, Boise folio, Idaho: U.S. Geological Survey Geologic Atlas, Folio 45, 7 p.

Meinzer, O. E., 1923a, The occurrence of ground water in the United States, with a discussion of principles: U.S. Geological Survey Water-Supply Paper 489, 321 p.

——1923b, Outline of ground-water hydrology, with definitions: U.S. Geological Survey Water-Supply Paper 494, 71 p.

Newell, F. H., and others, 1899, Report of progress of stream measurements for the calendar year 1897: U.S. Geological Survey 19th Annual Report, pt. IV, p. 1–632.

Palissy, Bernard, 1957, The admirable discourses, translated by A. La Rocque: University of Illinois Press, 264 p.

Russell, I. C., 1897, A reconnaissance in southeastern Washington: U.S. Geological Survey Water-Supply Paper 4, 96 p.

——1901, Geology and water resources of Nez Perce County, Idaho: U.S. Geological Survey Water-Supply Papers 53 and 54, 141 p.

——1903, Preliminary report on the artesian basin in southwestern Idaho and southeastern Oregon: U.S. Geological Survey Water-Supply Paper 78, 53 p.

Walcott, Charles D., 1896, Report of the Director: U.S. Geological Survey 17th Annual Report, pt. I—Director's Report and other papers, p. 5–207.

MANUSCRIPT RECEIVED BY THE SOCIETY SEPTEMBER 10, 1979
REVISED MANUSCRIPT RECEIVED JANUARY 8, 1980
MANUSCRIPT ACCEPTED MAY 20, 1980

Geological Society of America
Special Paper 183
1980

Gilbert—Bedding rhythms and geochronology

ALFRED G. FISCHER
Department of Geological and Geophysical Sciences, Princeton University, Princeton, New Jersey 08544

ABSTRACT

Gilbert interpreted rhythmic spacing in limestone units in the Upper Cretaceous of Colorado as a geological response to a planetary cause—the 21,000 yr precession of the equinoxes. In this light, he judged the Upper Cretaceous marine sequence of Colorado to have been deposited over a span of about 21 m.y.—a figure that seems remarkably close to that yielded by modern radiometric geochronology, 24 to 35 m.y.

A reexamination of three of Gilbert's four short rhythmic sequences, using the available radiometric data of Obradovich and Cobban, in conjunction with a model of subsidence and sedimentation, yields bedding rhythms in the 18,000 to 22,000-yr range, and seems to confirm Gilbert's hypothesis. Most rhythmic Cretaceous sequences in other parts of the world also yield bedding rhythms close to the precessional period, according to the Obradovich-Cobban time scale. The alternative Van Hinte time scale, however, yields a wider scatter of values, and suggests that only some of the rhythms are related to the precession, others seeming to be closer to the 41,000-yr period of obliquity.

The equinoctial precession can affect geology only when acting in conjunction with the eccentricity of the Earth's orbit, which waxes and wanes irregularly with a mean period of 93,000 yr. One should therefore expect precessionally caused rhythms to occur in sets averaging 4.5. That limestone-shale bedding rhythms occur in sets was shown long ago by Schwarzacher for upper Paleozoic and Mesozoic sequences. His observation that the mean number of rhythms per set lies between 5 and 6 suggests that the orbital parameters may have changed. At least two of Gilbert's four Cretaceous sequences of Colorado are bundled in this fashion.

Thus, Gilbert's suggestion that bedding rhythms provide a basis for geochronology takes on new interest—not to compete with radiometry in the rough calibration of Earth history, but as a refinement. It may also provide a means of tracing the evolution of the Earth's orbital behavior.

INTRODUCTION

In 1895, G. K. Gilbert wrote a short, pithy paper entitled "Sedimentary Measurement of Geological Time." It began with observations on rhythmic features in certain Cretaceous sediments near Pueblo, Colorado, and ended with a remarkably good estimate of the time represented by the marine Upper Cretaceous in the Western Interior Seaway. This paper and Gilbert's Presidential Address of 1900 to the American Association for the Advancement of Science shed a special light on the genius of this man. I attempt here to view this work in the perspective of 85 yr—a time that brought, among many other advances, that of radiometric dating, of a better base for interpreting the origins of sedimentary rocks, and of parallel observations in rocks of different ages and in other parts of the world.

GILBERT'S CONCEPTS

Rhythmic Sedimentation in the Cretaceous Rocks of Colorado

During the Cretaceous Colorado was a part of a great tectonic basin that linked the Arctic Ocean with the region of the Gulf of Mexico and the Caribbean, which with the Tethyan belt, formed, at times, a continuous marine strait between these oceanic regions. This basin was asymmetrical. Its western side formed a rapidly subsiding foredeep of molasse type, fronting the rising chains of what we now term the Sevier orogenic belt. In contrast, its eastern side, facing the stable craton, subsided more slowly. Whereas the western margin was primarily the site of alluvial and deltaic sedimentation, the central and eastern parts of the belt were characterized by the deposition of marine shales and at times carbonates. In the vicinity of Pueblo, Colorado, some 4,000 ft of such shales and minor carbonates intervene between the transgressive early Cenomanian Dakota Sandstone and the regressive late Maastrichtian Fox Hills Sandstone, and they were Gilbert's target.

Within that sequence, Gilbert's attention was caught by four short intervals in which limestone beds alternate rhythmically with interbeds of shale. In stratigraphic order, these sequences are (1) the Bridge Creek Member of the Greenhorn Formation, spanning the Cenomanian-Turonian boundary (Figs. 1, 2); (2) the Fort Hays Limestone Member of the Niobrara Formation, of Early Coniacian age (Fig. 3); (3) the "lower limestone" unit of the Smoky Hill Member of the Niobrara Formation, of late Coniacian age; (4) the uppermost part of the Smoky Hill Member of the Niobrara Formation, of early Campanian age.

Gilbert viewed the fine-grained limestones as chemical precipitates. The alternation of such chemical sediments with detrital

Figure 1. Bridge Creek Member, Greenhorn Formation. Railroad cut north of Arkansas River Dam near Pueblo, Colorado (Rock Creek anticline).

shale seemed to him not readily explained by the local vagaries of sorting and transport that produce an interbedding of shale and sandstone in the processes of detrital sedimentation; rather, they seemed to reflect an alternation of fundamentally different sedimentary regimes, likely to have affected much larger regions, and to have acted in response to regional if not global causes—*dictators* in the later terminology of Sander (1936, 1951). Furthermore, the spatial rhythm ("Raumrhythmus" of Sander) suggested the likelihood that these causes operated in a temporally rhythmical manner ("Zeitrhythmus" of Sander).

Gilbert's Search for a Cause

In searching for a possible cause, Gilbert turned first to the annual cycle: could these bedding couplets be what we have later come to know as *varves*? Gilbert dismissed that possibility convincingly. The mean thickness of couplets, in his four sequences, ranged from 1.5 to 2.6 ft. To supply such quantities of sediment annually to a great seaway called for roughly equivalent erosion rates in the source area, presumed to have lain to the west. Such erosion rates were several orders of magnitude larger than any that seemed geologically reasonable to Gilbert, who had earlier concerned himself with such matters. He was therefore driven to look for rhythms several orders of magnitude longer than the annual one. That brought him directly to the cycles of the Earth's orbital perturbations.

The existence of three kinds of long-range perturbations in the

Earth's orbit had been recognized, through centuries of astronomical observations (see Zeuner, 1952; Imbrie and Imbrie, 1979): the *precessional cycle* of the Earth's axis, which has a period of about 26,000 yr, the variation in the tilt of the Earth's axis, generally referred to as the *obliquity* or the angle of the ecliptic, having a period of about 41,000 yr; and a variation in the *eccentricity* of the Earth's orbit, with a somewhat variable amount and timing, but generally thought to average about 93,000 yr.

In theory, the precessional cycle should have no direct effect on Earth climates, except insofar as it interacts with the cycle in eccentricity. Because the ellipse itself moves in space, the *effective* period of the precession relative to that of the orbit has a mean length of 21,000 yr.

These perturbations had already been used by Croll (1875) and others to explain the rhythmic advance and retreat of the ice sheets during the Pleistocene. Blytt (1883, 1886, 1889) had suggested that the precessional cycle, acting through climatic agencies, had caused alternations between detrital and chemical sediments in the stratigraphic record.

Gilbert dismissed the eccentricity as being too weak and too irregular as a cause of his bedding rhythms. He ignored the cycle in obliquity, whose period remained unknown until later. Thus, he was left with the precessional cycle (in terms of the eccentricity) as the dictator.

Croll and Blytt had already suggested that the precessional cycle, in conjunction with an eccentric orbit, would affect the Earth's climates and thereby sea level, and in later times Milankovitch

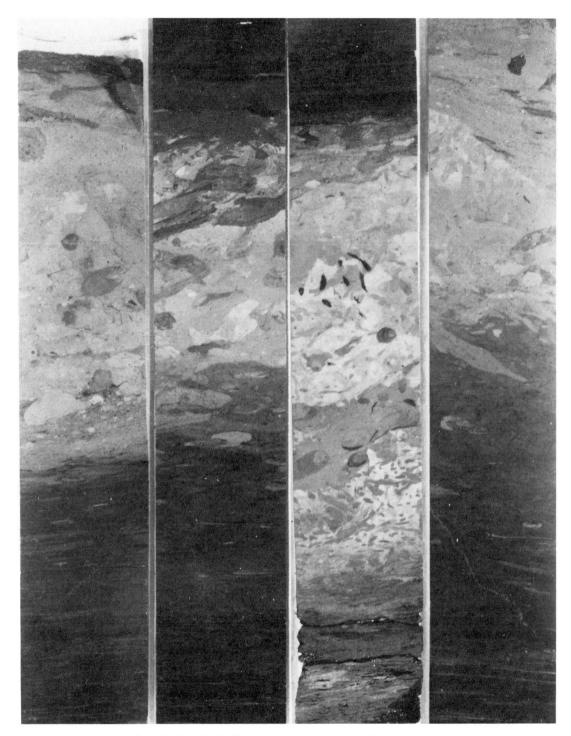

Figure 2. Four limestone beds of the Bridge Creek Member, and the associated bituminous shale and marl. Limestone beds are strongly burrowed by benthic fauna; shale and marl are laminated and undisturbed, except below limestone beds. Cores from Princeton–U.S. National Museum borehole below Arkansas River Dam, Pueblo, Colorado. Width of cores, 46 mm.

(1941) was to establish this more firmly. Gilbert considered several ways in which climatic changes might have caused the sedimentary alternations observed.

1. Climatic change might have caused glaciers to grow and shrink, raising and lowering Cretaceous sea levels in Colorado (though by Adhemar's mechanism rather than by the simple transfer of water from oceans to ice caps, which we invoke today).

Transgressions would bring carbonate deposition, regressions clay.

2. Global wind patterns might have changed, and with them currents in the Western Interior Seaway, so as to bring mud from the western shore at some times and not at others.

3. The climates of the western source lands might have oscillated between humid and arid. Dry times, with little plant cover and

Figure 3, Fort Hays Member, Niobrara Formation. At bridge of Route 96 across Peck Creek, west of Arkansas River Dam near Pueblo, Colorado.

strong runoff, would be likely to bring larger quantities of terrigenous matter to the seaway, over the years, than would humid times. The latter would favor a retention of soils, and more mature weathering, and would therefore bring more dissolved matter to the sea, favoring chemical over physical deposition. Gilbert, a strong believer in multiple hypotheses, did not express any preference for one or another of theses possibilities.

Gilbert's Geochronology

Throughout the nineteenth century, the time dimension of Earth history was a matter of highly divergent views. By Gilbert's day the biblical time scale of Bishop Usher was no longer taken seriously in scientific circles, but there remained wide discrepancies in opinion: Lyell (1867), using Croll's estimation of the length of the Pleistocene as 1 m.y. extrapolated evolutionary rates, to arrive at 20 m.y. for the length of the Neogene, 80 m.y. for the length of the Cenozoic, and 240 m.y. for the length of the Phanerozoic. Lord Kelvin, on the other hand, progressively refined his calculations on the age of a habitable Earth, from his earlier date of not more than 400 m.y. to a span of 20 to 40 m.y. (1899). Walcott (1893) used supposed depositional rates to calculate the length of the Phanerozoic at 27.5 m.y., and Blytt (1889) had calculated the length of the Cenozoic at 3.25 m.y., on the basis of transgressions equated with the cycle in orbital eccentricity.

Gilbert's approach allowed him first of all to attach time values to the four small rhythmic segments of the Cretaceous marine sequence. The longest of these contains no more than 33 couplets, equivalent to as many 21,000-yr cycles, or 693,000 yr. In ex-

trapolating this slender base to the entire sequence, he introduced a correction: Noting that the most calcareous parts of the sequence showed the thinnest couplets, with a mean of 1.5 ft, and the more shaley ones averaged 2.6 ft in thickness, he argued that the noncalcareous bulk of the sequence must have been deposited at yet greater rates, and suggested for the whole sequence a mean depositional rate of 4 ft per precessional cycle. Thus, for the whole 4,000-ft sequence, he allowed a time of 1,000 cycles, or 21 m.y. To this bold guess he attached error limits of a factor of +2 or -2; the sequence could represent no less than 10 and no more than 40 m.y.

WHAT HAVE WE LEARNED?

In turning from Gilbert's views of 1895 to the present, I shall reverse the sequence, to deal first with his general geochronology and second with the rhythmic phenomena on which he based it.

General Geochronology

Radioactivity was being discovered at the time of Gilbert's work. Within a few years the first radiometric age determinations were made, and from then on geochronology became virtually synonymous with the geochemical approach. Two generations of such work have now brought us to the point at which the lengths of the various periods are known with reasonable confidence to within 5% to 10% of their value, and the lengths of most stages may be known to within a factor of 2. This has been extremely important to geology, but as any stratigrapher knows, it has not established an accurate chronology at the million-year level for any part of geological time excepting the latter part of the Neogene.

This problem is exemplified by the current existence of two rival radiometric time scales for the Cretaceous (Table 1). One has been proposed by Obradovich and Cobban (1975), based on ash beds in the deposits of the Western Interior Seaway. For the sake of simplicity, we are here using the slight variation of this scale introduced by Kauffman (1977). The other scale is that of Van Hinte (1976), based largely on glauconite from European sequences. Both time scales are based on many analyses, from beds that are members of the stratigraphic succession but both suffer from the inaccuracies inherent in dates derived from glauconite and from the devitrified ashes: The scatter of dates derived from a single stratigraphic member that was deposited over a span of less than a million years may be several million years (Fig.5). The dates thus serve only as general guidelines, and between them the investigator is forced to interpolate by reference to sedimentation rates or number of paleontological zones.

Figure 4 is a Bubnoff or geohistory diagram (Fischer, 1964, 1975; Van Hinte, 1978) for the shaley part of the Upper Cretaceous stratigraphic sequence in the Pueblo region. This plot begins at the base of the Graneros Shale (early Cenomanian) and ends at the top of the Pierre Shale of Maastrichtian age, which is here arbitrarily normalized to an age of 68 m.y.B.P. The circles represent the depth to which the base of the Graneros Shale had subsided below sea level, at times corresponding to the ages of boundaries between the succeeding stratigraphic members or formations (see Fig. 5). No corrections have been made for water depth or compaction.

The open circles (curve A–A′–E) are plotted according to chronology of Van Hinte; the solid circles (curve B–B′–E) are according to that of Obradovich and Cobban modified by Kauffman (1977). Both show a progressive increase for subsidence and sedimentation rates and thus differ from the linear sedimentation envisioned by Gilbert (curve C–E). Nevertheless, Gilbert's time for the sequence (21 m.y.) is not far out of line, and the mean rates of his limits of confidence (21 and 116 Bubnoff units) comfortably include the mean rates of the Obradovich-Cobban and Van Hinte models (44 and 38 Bubnoff units, respectively).

The degree to which Gilbert's stab at geological time anticipated the radiometric results is surprising: Was he then right in ascribing a 21,000-yr rhythm to bedding couplets? A few geologists of the next two generations were inclined to follow him, but the majority of sedimentologists and stratigraphers ignored this suggestion.

Sedimentation of Cretaceous Beds—General

Part of the Cretaceous stratigraphy in the vicinity of Pueblo, Colorado, has been described by Cobban and Scott (1972). The nature and interpretation of the Cretaceous sediments of the Western Interior Basin, with special reference to the Colorado-Kansas sector, has been summarized by Kauffman (1977). The long transcontinental basin varied in character through Cretaceous time, from a fluviatile plain at one extreme to that of a marine seaway bordered in the west by deltas. Most of the time, however, the seas that occupied the region seem to have deviated from normal marine character (Scholle and Arthur, 1979): The peculiarly restricted faunas, largely molluscan, as well as abnormal values in oxygen isotopes, suggest that the seas were not of normal salinity. Furthermore, the great prevalence of black, carbon-rich shales, lacking burrows and containing either no benthic fossils or

TABLE 1. COMPARISON OF CRETACEOUS TIME SCALES

Stage	Obradovich-Cobban Yr B.P.	Kauffman variation Yr B.P.	Kauffman variation Duration	Van Hinte Yr B.P.	Van Hinte Duration
	64–65	64.00		65.00	
Maastricht.			4.75		5.00
	70–71	68.75		70.00	
Campanian			13.50		8.00
	82+	82.25		78.00	
Santonian			3.25		4.00
	86+	85.50		82.00	
Coniacian			1.50		4.00
	87+	87.00		86.00	
Turonian			4.00		6.00
	89–90	91.00		92.00	
Cenoman.			2.50		8.00
	94+	93.50		100.00	

only an extremely meager epifauna, suggests that the sea floor was very prone to stagnation—a condition that develops in such marginal seas as the Baltic and the Black Sea as the result of a strong salinity stratification.

Excursions of the depositional system through parts or all of this marine-continental spectrum of depositional environments have allowed Kauffman (1967, 1977) to recognize 10 Cretaceous cyclothems. Gilbert's rhythmic sequences represent the peak marine conditions attained in the Western Interior Seaway—specifically, the peaks of the Greenhorn cyclothem, T-6, and the Niobrara cyclothem, T-7, of Kauffman (1977).

Character and Origin of Rhythmic Deposits

The four sequences of rhythmic strata studies by Gilbert are in ascending order as follows (Scott and Cobban, 1974; Kauffman, 1977): (1) the Bridge Creek Member of the Greenhorn Formation, of latest Cenomanian–early Turonian age: 15.9 m thick, predominantly of dark, orgainc-rich, laminated shale, but containing about 30 beds of limestone bearing a series of rich marine faunas (Figs. 1, 2). (2) the Fort Hays Limestone Member of the Niobrara Formation of early Coniacian age: 12.2 m of limestone, consisting of about 33 beds separated by thin shales or shaley bedding planes (Fig. 3); (3) the "lower limestone unit" within the Smoky Hill Member of the Niobrara Formation of late Coniacian age: 11.6 m of interbedded shale and limestone beds, forming 16 couplets; and (4) a thin unit of rhythmically interbedded shale and chalk at the top of the Niobrara Formation, which is not as well studied as the other units and will not be dealt with here.

These rhythmic intervals represent the closest approach to normal marine conditions attained in this part of the seaway, times when it was most open to the ocean. The Bridge Creek Member is the peak of the Greenhorn transgression, whereas the successive rhythmic units are minor peaks of the Niobrara transgression.

The limestone beds are not of chemical origin, as Gilbert seems to have believed, but are biogenic (Hattin, 1971, 1975). They represent mixtures of coccolith-foraminiferal ooze with shells and shell debris of pelagic and benthic invertebrates. Inoceramids generally play a particularly prominent role in them, and echinoids have been reported from some. These limestone beds are intensely burrow-mottled (Fig. 2). They are thus shown to represent episodes of essentially normal marine conditions and generally aerated sea floors.

The associated shaley beds are finely laminated, largely unburrowed, generally bitumen-rich sequences of shale and coccolith or coccolith-globigerinid marl (Fig. 2). Evidence of benthic habitation is restricted to scattered millimetre-thick layers of

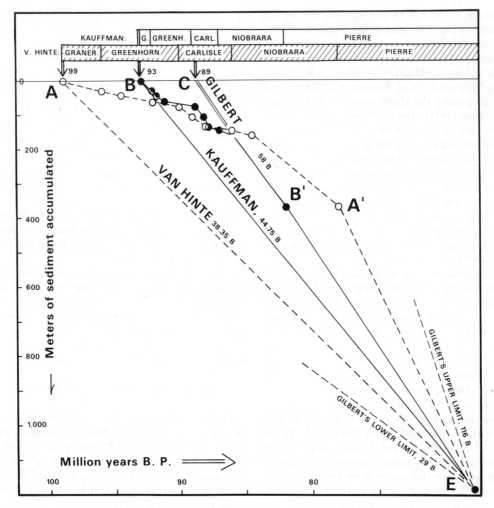

Figure 4. Three models for the accumulation of the upper Cretaceous shaley sequence at Pueblo, Colorado, comparing the chronologies of Van Hinte and of Kauffman and the timing proposed by Gilbert.

benthic shell debris and incipient burrow-horizons. These shaley intervals thus represent episodes during which the bottoms were generally in a dysaerobic to anaerobic state, not available to colonization by benthic shell-producing and/or burrowing organisms, save during brief exceptional periods. However, inoceramid debris, present within these shales in scattered fragments and in coprolites, suggests that even at these times local patches of the bottom were colonized. These served as centers from which debris was dispersed over dead bottoms by wide-ranging fishes and/or reptiles.

The oscillation between shale and limestone, within these intervals, reflects an alternation between times in which the water column was generally highly stratified and times in which it was generally well mixed. This alternation could be due to several causes, essentially those discussed by Gilbert, though operating in a slightly different manner.

A Minor Oscillation in Sea Level. Superimposed on the large-scale eustatic fluctuations that produced the Greenhorn and Niobrara transgressions, minor eustatic highs opened the seaway to inflow of ocean currents, while lows in oscillation imposed returns to the salinity-stratified state even at peak transgression on the longer cycle. Episodic glaciation remains a possible cause of such oscillations.

Changes in Precipitation. A change in climate brought a marked change in the amount of fresh water contributed to the seaway. Wet times were conducive to the development of haloclines and stagnation; dry times were not.

Winds and Current Regimes. The direction and intensity of winds and currents oscillated between two regimes, one of which allowed the water column to become more or less permanently stratified, whereas the other did not.

Periods of Gilbert's Rhythms

The first question that arises is whether the existence of a spatial rhythm in sediments is the result of a process operating rhythmically in time. Some sedimentologists (Duff and others, 1967) were inclined to be skeptical about this, whereas others such as Gilbert and Sander have been more positive. So long as the possibility of comparing the length of individual couplets by some independent geochronology remains completely out of reach, the choice is essentially a philosophical one: The world is clearly governed by a mixture of predictable and of stochastic processes, and one may lean toward one side or the other. In this case, Gilbert chose order over disorder and proposed a known cyclic cause, and we shall proceed with him to assume that the periods of succeeding

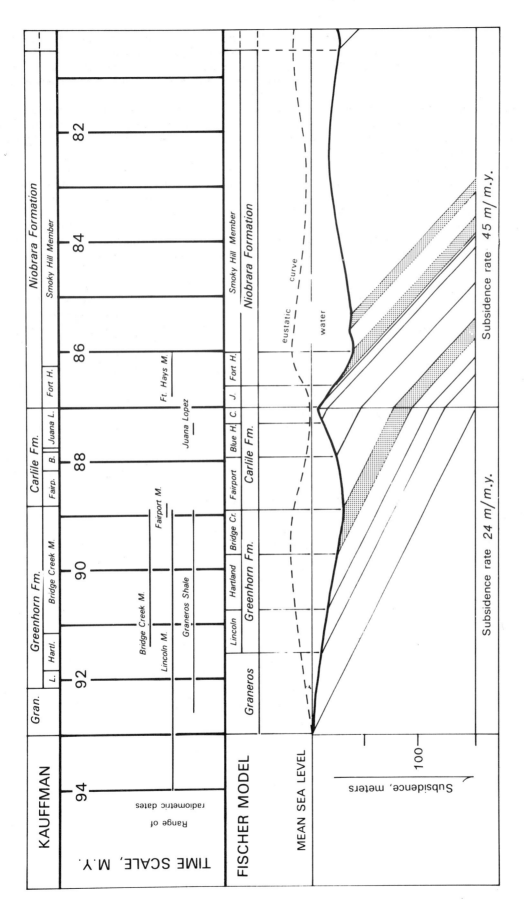

Figure 5. Two chronological models for Cretaceous events at Pueblo, Colorado. Upper half of diagram: time scale and radiometric data for the various members of the upper Cretaceous sequence; at very top, the chronology proposed by Kauffman. Lower half of diagram: subsidence diagram combining tectonic subsidence, eustatic oscillations, and bathymetry with a less literal match to the radiometric data, to provide an alternative chronological model. The rhythmically bedded units are shaded.

sedimentary couplets are comparable. The attempt will be to discover the length of this period of determining the duration of a rhythmic sequence of beds, and by apportioning that time equally between the contained couplets.

The general chronology of the Late Cretaceous has been discussed above (Fig. 4). In order to evaluate the timing of bedding couplets in the lower three of Gilbert's four rhythmic units, we must examine more closely the radiometric data available for that part of the section—from the ash falls within the Cenomanian, Turonian, and Coniacian stages of the Western Interior Basin. These dates are recorded by Kauffman (1977) and are plotted on Figure 5. Individual ashes have provided dates ranging through several million years, and dates from the Lincoln Member, for example, completely overlap those from the underlying Graneros Shale and from the overlying units (Bridge Creek Member). These are not time-transgressive sequences, time-lines being well established by correlated ash falls (Hattin, 1971), and thus it is clear that Obradovich and Cobban's caution about the meaning of individual dates must be heeded: The ages and durations of the various stratigraphic members are not sharply defined.

Kauffman (1977) has estimated them by averaging the dates (top of Fig. 5). This provides a surprising discrepancy (Table 2): The Bridge Creek Member, a black shale–marl sequence containing rhythmic limestones, yields an anomalously low sedimentation rate as compared to the black shaley units above and below and compared to the rhythmic limestone units higher in the section.

The lower half of Figure 5 represents an alternative approach to establishing dates and durations for the various stratigraphic members by treating them as successive members of a historical chain of events that involved tectonic subsidence, eustatic fluctuations, and changing supply of sediments. For the overall timing of this history, I have accepted Kauffman's dates as 93 m.y.B.P. for the base of the Graneros Shale, and 80.4 m.y. as the end of Niobrara time. During this stretch of time, there occurred two eustatic oscillations, the first of which corresponds to the Greenhorn cyclothem, or T-6 of Kauffman, the second to the first part of the Niobrara cyclothem, or T-7. These oscillations, shown by the dotted line, are here estimated at 20 m.

The first of these cycles actually began with the deposition of the largely fluviatile Dakota Sandstone (below the Graneros Shale and off the left end of the diagram), and passed through a shoreline

phase at 93 m.y.B.P. in the latest beds of the Dakota Formation (or base Graneros Shale). Peak marine conditions were reached in Bridge Creek time, and conditions returned to a brackish shoal-water state in the Codell Sandstone (C), whose age is estimated by both Kauffman and me at close to 87 m.y.

The question now is how the 6 m.y. of this history are to be apportioned to the various members; Kauffman has done it by averaging dates, and I have done it by assuming a steady rate of subsidence and what I consider to be a reasonable bathymetric history. The sediments that accumulated during this 6 m.y. span total 147 m. Inasmuch as this interval commenced and ended in the shoal-water state, these 147 m record the subsidence relative to sea level, uncorrected for the effects of compaction. This converts to a mean subsidence rate of 24 Bubnoff units (metres per million years, Fischer, 1969). If we assume that this was the actual rate of subsidence, each stratigraphic contact describes a linear path on the diagram, from its inception at the depositional interface to its position below the top of the Codell Sandstone at 87 m.y.B.P. Thus, on such a diagram, the timing of the stratigraphic boundaries is fixed by the intersection of such lines and the presumed bathymetry. For this bathymetry, I have assumed a maximum of 50m, attained gradually and lost somewhat more rapidly toward the end of the cycle. Bathymetry can be adjusted within limits to provide a reasonable match with the radiometric dates. Eicher (1969) assumed a maximal water depth of 500 m, whereas Kauffman (1967) suggested it lay between 30 and 150 m.

I make no claim that this approach leads to accurate age determinations, but it does offer some additional constraints beyond those of radiometry. In this case, it suggests that time was more evenly distributed between the Graneros Shale and the various members of the Greenhorn Formation than was suggested by Kauffman; specifically, that the Bridge Creek Member may only be a third as long as Kauffman suggested.

The next segment of time, the Niobrara cycle, extends from the shoal-water stage of the Juana Lopez Member to that of the Apache Creek Sandstone Member of the Pierre Formation, off the right end of the diagram (78.5 m.y.B.P. according to Kauffman, 1977). Here the subsidence rate was higher—on the order of 45 Bubnoff units, as is consistent with the overall picture of Cretaceous history (Fig. 4). A similar construction is used to date the members, and yields results which are not at variance with Kauffman's estimates.

Table 2 compares the apparent durations and sedimentation rates obtained for the various members by Kauffman with those obtained from the subsidence model of Figure 5. The mean period of bedding couplets in the rhythmic members is obtained by dividing the length of time by the number of couplets, and Table 3

TABLE 2. DURATION AND SEDIMENTATION RATE OF LITHOSTRATIGRAPHIC UNITS, UPPER CRETACEOUS, PUEBLO, COLORADO

	Duration (m.y.)		Sedimentation rate (m/m.y.)	
	Kauffman	Fischer	Kauffman	Fischer
Smoky Hill M.	4.75	5.60	44.7	37.9
Fort Hays M.	0.75	0.50	16.3	24.4
Juana Lopez M.	0.75	0 40	0.8	1.5
Codell M.	0.15	0.25	61.3	36.8
Blue Hills M.	0.30	0.70	102.6	44.0
Fairport M.	0.60	1.00	50.3	30.2
Bridge Creek M.	2.45	0.80	6.5	19.8
Hartland M.	0.60	1.00	30.0	18.0
Lincoln M.	0.40	0.80	45.6	14.2
Graneros Fm.	0.50	1.50	62.8	20.9

TABLE 3. PERIODS OF BEDDING RHYTHMS IN LATE CRETACEOUS AT PUEBLO, COLORADO

	Kauffman	Fischer
"Lower limestone" 16 couplets		19,000
Fort Hays Member 33 couplets	18,000 – 20,000	18,000
Bridge Creek Member 30 couplets	60,000 – 80,000	27,000
Lower 2/3 of Bridge Creek Member, 24 couplets		22,000

Note: According to chronologies of Kauffman and of Fischer (Fig. 5).

compares the results. For the Bridge Creek Member, these are discrepant: The subsidence model yields values close to the precessional period, whereas Kauffman's approach yields a period three times as long, somewhere between that of the cycle in obliquity and the cycle in eccentricity. The Fort Hays Member yields a period in the precessional range by either approach, and the period of the "lower limestone" cycles seems precessional as well.

We thus conclude that such radiometric data as are now available do not provide either proof or disproof for Gilbert's hypothesis, but that in two out of three instances they offer strong support for it; the third instance (Bridge Creek Member) remains equivocal.

Rhythmicity in the Late Cretaceous of Italy

Sedimentary sequences in which shale alternates rhythmically with limestone are widespread in the geological record (Schwarzacher, 1975). For some years I have beeen involved with others in a study of a pelagic sedimentary sequence of this sort, the Scaglia Formation, at Gubbio in the Umbrian Appennines of Italy (Arthur and Fischer, 1977). The record appears to be an essentially continuous one, containing all of the recognized Cretaceous and Paleocene foraminiferal zones, as well as magnetic zones that match the record of the sea-floor anomalies (Alvarez and others, 1977). We shall here limit our discussion to bedding rhythms of the Upper Cretaceous. The stage boundaries used are those of Premoli Silva and others (1977) and are based primarily on planktonic foraminifera.

Arthur (1979) measured bed-by-bed sections for fractions of five stages. Extrapolating these results to stage length, he arrived at mean periods of bedding couplets, using the stage durations of the Van Hinte scale (Table 4). Also shown on Table 4 are calculations made according to the Obradovich-Cobban scale. As noted in Table 1, these scales are in essential agreement for the lengths of some stages, but differ for others by as much as a factor of 3. The differences may lie partly in problems of correlation: Time correlations between the sediments of the North American interior (on which the Obradovich-Cobban scale is based) with the classical sections of western Europe leave something to be desired. However, the main reason for the discrepancies probably lies in the shortcomings of glauconite (in Europe) and of volcanic ashes (in North America) as source materials for radiometric chronology.

As shown on Table 4, the periods of couplets when viewed on the Obradovich-Cobban scale yield values ranging from 10,000 to 35,000 yr, and *averaging 20,000 yr*. On the Van Hinte scale, they

TABLE 4. PERIODS OF BEDDING RHYTHMS IN LATE CRETACEOUS, SCAGLIA FORMATION, GUBBIO, ITALY

Stage	Obradovich and Cobban	Van Hinte
Maastrichtian	15,000	16,000
Campanian	30,000	18,000
Coniacian	10,000	27,000
Turonian	35,000	49,000
Cenomanian	15,000	49,000
Mean period	20,000	32,000

Note: From measurements by Arthur (1979), calculated according to the time scales of Obradovich and Cobban and Van Hinte.

range between 16,000 and 49,000 yr and *average 32,000 yr*. The Obradovich-Cobban scale thus offers strong support for Gilbert's view that the bedding couplets are tied to the precession. The Van Hinte Scale on the other hand suggests that if they are controlled by orbital parameters, some are precessional and others are perhaps related to the cycle in obliquity.

Thus, the application of radiometric data to bedding rhythms in the Upper Cretaceous parts of the Scaglia Formation at Gubbio leads to results similar to those obtained for Colorado: The bedding rhythms lie in the general range of the periods of orbital perturbations. Under one set of assumptions, the mean period of bedding couplets is essentially that of the precession, 21,000 yr. Under another set of assumptions, it is longer and suggests that the sequence contains portions in which bedding couplets are related to the precession, and other portions in which they reflect some cycle of a longer period—most likely that of obliquity.

Bundled Rhythms

As Zeuner and others (1952) have pointed out, the precessional rhythm can influence Earth climate only by way of the eccentricity of the orbit. Because the latter is cyclic, with a mean period about four or five times that of the precessional cycle, it follows that precessionally induced climatic fluctuations should wax and wane in a rhythmic pattern. Accordingly, bedding rhythms, if an expression of such climatic fluctuations, might be expected to occur in bundles or sets.

Such a bundling in bedding rhythms has long been recognized by Schwarzacher who has found it in the Alpine Triassic (1954), as well as in the Carboniferous of Ireland and the Jurassic of Germany (1975). The number of bedding couplets in a bundle varies, but averages near 5, an observation that we may term "Schwarzacher's Rule." It applies also to the Triassic of the Newark basin (Van Houten, 1964).

Turning back to the Cretaceous of Colorado, we note that the bedding couplets in the Fort Hays Limestone show such a bundling (Fig. 6). Delimitation of bundles is not clean and unequivocal; bundles defined by thickness of shaley interbeds contain an average of 4.4 bedding couplets, whereas bundles defined by bedding thickness of limestone average 6. The third of Gilbert's units (lower limestone) is also bundled, whereas the Bridge Creek couplets (Fig. 1) are not. Thus, this line of evidence also suggests that the Fort Hays and "lower limestone" couplets are of precessional origin, and leaves the origin of the Bridge Creek couplets in doubt.

Comparison with the Pleistocene

Pleistocene stratigraphers have been more inclined than others to attribute cyclical phenomena to orbital causes, ever since Croll (1875). Of particular interest is a paper by Hays and others (1976), in which the variations in two kinds of faunal indices and oxygen isotope ratios, all presumably correlated directly or indirectly with climatic factors, are analyzed statistically within two particularly fine deep-sea cores. The spectral frequencies in these fluctuations cluster around three periods: a multimodal one in the 20,000-yr range, presumably precessional; one of around 40,000 yr, and presumably related to the obliquity cycle; and one that lies near 100,000 yr, presumably reflecting eccentricity.

Thus, orbital perturbations are seen reflected in ice regimes and via this phenomenon in the record of marine sediments. The Late

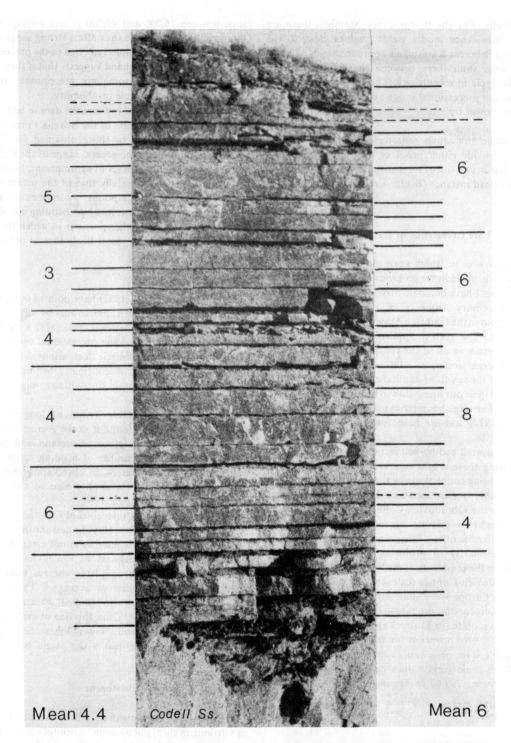

5

3

4

4

6

Mean 4.4

6

6

8

4

Mean 6

Codell Ss.

Figure 6. Fort Hays Member, Niobrara Formation, enlargement of Figure 2 to show bundling of bedding couplets and alternatives in their interpretation. At left, bundles defined by major shaley interbeds, averaging 4.4 couplets per bundle. At right, bundles defined by variations in thickness of limestone beds, averaging 6 couplets per bundle.

Cretaceous record suggests that sedimentary patterns reflect orbital perturbations also at times at which ice volume was either very small or nonexistent.

CONCLUSIONS

Gilbert suggested in 1895 that rhythmic bedding phenomena have resulted from climatic responses to orbital perturbations, specifically to the equinoctial precession, and thus provide a 21,000-yr yardstick for the measurement of geologic time intervals.

As yet it has not been possible to test this proposition rigorously by means of the radiometric time scale, due chiefly to the inaccuracies within the latter and to the difficulty of applying it to specific and limited parts of stratigraphic sequences.

Nevertheless, the bulk of the evidence accumulated since Gilbert's day supports the following generalizations:

1. Rhythmic sequences of limestone-shale bedding couplets in the Late Cretaceous show mean periodicities that fall into the range of the Earth's orbital perturbations.

2. When viewed in the light of the Obradovich-Cobban time scale, all of the Upper Cretaceous sequences studied by me to date suggest a precessional period, supporting Gilbert (note, however, the divergence of Kauffman's model).

3. When viewed in the light of the Van Hinte time scale, this generalization breaks down; some couplet sequences appear precessional, others have longer periods (perhaps the 41,000-yr cycle of obliquity).

4. Further evidence for the precessional nature of some couplet sequences is seen in their bundling into sets of ± 5(Schwarzacher's Rule), the bundle representing the period of the cycle in eccentricity. Indeed, perhaps one should be skeptical about the precessional nature of bedding couplets that are not bundled.

A geochronology based on orbital rhythms, such as visualized by Gilbert, is not as yet operational. At best, it will be limited to rhythmically structured sequences in which secondary overprints such as intraformational erosion or the amalgamating effects of bioturbation can be discounted. Important questions such as changes in the orbital perturbation parameters through time, and changes in the Earth's response to them, remain to be explored.

These phenomena in themselves constitute first-order problems in historical geology. Furthermore, the development of a geochronology based on clocks that strike at intervals of 20,000 to 100,000 yr would provide approaches to the measurement of geological processes for which the radiometric approach has proved too coarse.

The seeds planted by Gilbert in 1895 have been largely dormant over the last 85 years, but may be about to sprout. Is it too much to hope that they will bear fruit before they reach their centennial? Whatever the outcome, the fact that Gilbert's shadow falls across nearly three generations, to play a role in some of the frontier questions of modern historical geology, demonstrates the stature of this extraordinary man.

ACKNOWLEDGMENTS

For help in field work and for laboratory data, I am indebted to Michael Arthur and to Lisa Pratt. Erle Kauffman and Donald Hattin kindly introduced me to the western Cretaceous. Travels in the course of a Guggenheim Fellowship first brought these rhythmic Cretaceous sequences to my attention, and later work on them has been supported largely through grant EAR 77-23369 of the National Science Foundation. Michael Arthur, Peter Scholle, Norman Sohl, and Ellis Yochelson kindly read the paper and suggested improvements.

REFERENCES CITED

Alvarez, W., Arthur, M. A., Fischer, A. G., Lowrie, W., Premoli Silva, I., and Roggenthen, W. M., 1977, Upper Cretaceous-Paleocene magnetic stratigraphy at Gubbio, Italy: V, Type section for the Late Cretaceous-Paleocene geomagnetic reversal time scale: Geological Society of America Bulletin, v. 88, p. 383-389.

Arthur, M. A., 1979, Sedimentologic and geochemical studies of Cretaceous and Paleogene pelagic sedimentary rocks: The Gubbio sequence. Part I [Ph.D. dissert.]: Princeton University, 174 p.

Arthur, M. A., and Fischer, A. G., 1977, Upper Cretaceous-Paleocene magnetic stratigraphy at Gubbio, Italy: I, Lithostratigraphy and sedimentology: Geological Society of American Bulletin, v. 88, p. 367-371.

Blytt, Axel, 1883, Om vexellagning og dens mulige betydning for tidsregningen i geologien og laeren om arteres forandringer: Videnskabs.-Selskabs. Forhandl., Christiania, 1883, no. 9, 31 p., 1 pl.

——1886, On variations of climate in the course of time: Videnskabs.-Selskabs. Forhandl. Christiania, 1886, no. 8, 24 p.

——1889, The probable cause of the displacement of beach-lines, an attempt to compute geological epochs: Videnskabs.-Selskabs. Forhandl. Christiania, no. 1, 93 p.

Cobban, W. A., and Scott, G. R., 1972, Stratigraphy and ammonite fauna of the Graneros Shale and Greenhorn Limestone near Pueblo, Colorado: U.S. Geological Survey Professional paper 645, 108 p., 41 pls.

Croll, J., 1875, Climate and time in their geological relations: A theory of secular change in the earth's climate: London, 577 p., 7 pls.

Duff, P. M., Hallam, A., and Walton, E., 1967, Cyclic sedimentation: Elsevier Publishing Company, 280 p.

Eicher, D. L., 1969, Paleobathymetry of Cretaceous Greenhorn sea in eastern Colorado: American Association of Petroleum Geologists Bulletin, v. 53, p. 1075-1090.

Fischer, A. G., 1964, The Lofer cyclothems of the Alpine Triassic, in Merriam, D. F., ed., symposium on cyclic sedimentation: Kansas State Geological Survey Bulletin 169, v. 1, p. 107-150.

——1969, Geological time-distance rates: The Bubnoff unit: Geological Society of America Bulletin, v. 86, p. 49-51.

——1975, Origin and growth of basins, in Fischer, A. G., and Judson, Sheldon, Petroleum and global tectonics: Princeton University Press, p. 47-79.

Gilbert, G. K., 1895, Sedimentary measurement of geologic time: Journal of Geology, v. 3, p. 121-127.

——1900, Rhythms and geologic time: American Association for the Advancement of Science Proceedings, v. 49, p. 1-19.

Hattin, D. E., 1971, Widespread, synchronously deposited, burrow-mottled limestone beds in Greenhorn Limestone (Upper Cretaceous) of Kansas and southeastern Colorado: American Association of Petroleum Geologists Bulletin, v. 55, p. 412-431.

——1975, Stratigraphy and depositional environments of the Greenhorn Limestone (Upper Cretaceous) of Kansas: Kansas Geological Survey Bulletin 209, 128 p.

Hays, J. D., Imbrie, J., and Shackleton, N. J., 1976, Variations in the earth's orbit: Pacemaker of the ice ages. Science, v. 194, p. 1121-1132.

Imbrie, John, and Imbrie, K. P., 1979, Ice ages, solving the mystery: Short Hills, New Jersey, Enslow Publishers, 224 p.

Kauffman, E. G., 1967, Coloradoan macroinvertebrate assemblages, Central Western Interior, United States; *in* A symposium on paleoenvironments of the Cretaceous seaway in the Western Interior: Golden, Colorado, Colorado School of Mines, p. 67–143.

———1977, Cretaceous facies, faunas, and paleoenvironments across the Western Interior Basin: The Mountain Geologist, v. 14, no. 2–3, p. 75–274.

Kelvin, Lord, 1899, The age of the earth as an abode fitted for life: Philosophical Magazine, v. 4, p. 66–90.

Lyell, Sir Ch., 1867, Principles of geology (10th edition, 2 volumes): London, J. Murray.

Milankovitch, M., 1941, Kanon der Erdbestrahlung and seine Anwendung auf das Eiszeitenproglem (special edition): Academie Royale Serbe, Belgrade, 133, 633 p.

Obradovich, J. D., and Cobban, W. A., 1975, A time scale for the Late Cretaceous of the Western Interior of North America, *in* Caldwell, W. G. E., ed., The Cretaceous system in the Western Interior of North America: Geological Association of Canada Special Paper 13, p. 31–54.

Premoli Silva, I., Paggi, L., and Monechi, S., 1977, Cretaceous through Paleocene biostratigraphy of the pelagic sequence at Gubbio, Italy, *in* Pialli, G., ed., Paleomagnetic stratigraphy of pelagic carbonate sediments: Memorie della Societa Geologica Italiana, v. 15, p. 21–32.

Sander, Bruno, 1936, Beiträge zur Kenntnis der Anlagerungsgefüge: Tschermaks Mineralogische und Petrographische Mitteilungen, v. 48, p. 27–139.

———1951, Contribution to the study of depositional fabrics: American Association of Petroleum Geologists, 160 p.

Scholle, P. A., and Arthur, M. A., 1979, Carbon isotopic fluctuation in pelagic limestones: potential stratigraphic and petroleum exploration tool: American Association of Petroleum Geologists Bulletin, v. 64, p. 67–87.

Schwarzacher, W., 1954, Die Grossrhythmik des Dachsteinkalkes von Lofer: Tschermaks Mineralogische und Petrographische Mitteilungen, ser. 3, v. 4, p. 44–54.

———1975, Sedimentation models and quantitative stratigraphy. Development in stratigraphy: Amsterdam, New York, Elsevier Publishing Company, v. 19, 377 p.

Scott, G. R., and Cobban, W. A., 1964, Stratigraphy of the Niobrara Formation at Pueblo, Colorado: U.S. Geological Survey Professional Paper 454–L, 30 p., 11 pls.

Van Hinte, J. E., 1976, A Cretaceous time scale: American Association of Petroleum Geologists Bulletin, v. 60, p. 498–516.

———1978, Geohistory analysis—Application of micropaleontology in exploration geology: American Association of Petroleum Geologists Bulletin, v. 62, p. 201–222.

Van Houten. F. B., 1964, Cyclic lacustrine sedimentation, Upper Triassic Lockatong Formation, central New Jersey and adjacent Pennsylvania: Kansas Geological Survey Bulletin 169, p. 497–531.

Walcott, C. D., 1893, Geologic time as indicated by the sedimentary rocks of North America: Journal of Geology, v. 1, p. 639–676.

Zeuner, F. O., 1952, Dating the past: London, Methuen and Company, 495 p.

MANUSCRIPT RECEIVED BY THE SOCIETY MAY 20, 1980
MANUSCRIPT ACCEPTED MAY 20, 1980

Geological Society of America
Special Paper 183
1980

Grove Karl Gilbert and the origin of barrier shorelines

JOHN C. KRAFT
Department of Geology, University of Delaware, Newark, Delaware 19711

ABSTRACT

In the latter part of the nineteenth century, G. K. Gilbert began a study of the origin of the lakeshore features of ancestral Lake Bonneville. By means of hypothesis and observation, he used features of the shorelines of the Great Lakes and the Atlantic and Pacific Oceans to form modern-ancient analogs. Studies of present coastal processes and the geometry and internal structure of the Lake Bonneville shorelines lead to the hypothesis that the littoral transport mechanism was dominant in the formation of lagoon-barrier coastal systems. His works on barrier evolution have stood the test of time. Although some of the world's barrier shorelines have evolved by other processes, most of them appear to fit Gilbert's hypothesis for barrier evolution. Furthermore, the process of littoral transport appears to be of great importance in modification or alteration of coastal barrier landforms no matter what their origin.

INTRODUCTION

Over the past century, geologists have entertained many hypotheses concerning the evolution of spits, barriers, and lagoons. Some of these hypotheses are based on geomorphology, and others are based on a combination of morphology, processes, and subsurface geology. G. K. Gilbert inferred vertical crustal movement from submergence of marsh plants and from the presence or absence of wave-cut cliffs. However, his main contribution to coastal geology concerns the development of barriers. Gilbert's contributions include his work "The Topographic Features of Lake Shores" (Gilbert, 1885), his monograph on Lake Bonneville (1890), and his supplementary discussions of coastal geology (Gilbert 1886, 1900, 1917). The impact of these studies on current knowledge about coastal barriers was and remains profound. In his paper of 1885, "The Topographic Features of Lake Shores," Gilbert presented a unique work in a new field with few earlier contributions. Only the works of de Beaumont (1845) on the evolution of barriers by the emergence of a bar from a shoaling of the sea bottom, Alesandro Cialdi (1866, pers. commun. to Gilbert, 1885, p. 766) on the motion of waves and their action on coasts, H. Keller (reference unknown) on the formation of sandy coasts in Germany, and Mitchell (1869) on coastal tide lands seemed important enough to Gilbert to cite in his paper.

Johnson (1919) presented the first detailed review of the evolution of barrier shorelines. Contrasting the work of G. K. Gilbert with that of de Beaumont, Johnson generally supported de Beaumont's hypothesis that barriers formed by the shoaling of offshore bars and upbuilding of the barriers from the sea bottom, but Johnson also pointed out several cases in which barriers might have originated by deposition of longshore drift, as advocated by G. K. Gilbert. Johnson (1919, p. 365) concluded: "It is perhaps more reasonable to accept the de Beaumont theory of bar formation, not forgetting, however, that longshore transportation of debris is an accessory process of very great importance." The failure of Johnson's system and his misidentification of the Atlantic Coastal Plain as one of emergence as opposed to submergence has led to a denigration of Johnson's efforts. However, they were fundamental to the entire problem and set the premise followed in many hundreds of subsequent papers.

The literature now available on the origin and classification of barriers is much too large to be reviewed in this paper. However, a handful of possible modes of origins of barrier islands can now be identified (Schwartz, 1971; Fisher, 1979). They include (1) the emergence of offshore bars in a stable or rising sea-level condition (de Beaumont, 1845), (2) littoral transportation of materials from an eroding shoreline across a reentrant as a spit tying to the next headland (Gilbert, 1885), (3) a rising sea level drowning shore-parallel features such as coastal dunes (McGee, 1890; Hoyt, 1967), and (4) combinations such as those suggested by Pierce and Colquhoun (1970) regarding the initial development of a primary barrier by coastal erosion against a land undergoing transgression followed by a secondary barrier offshore formed by 1 or 2, noted above. It is interesting to note that all the hypotheses require a strong element of littoral transportation (littoral transport) which is Gilbert's principal mechanism for the evolution and continuing development of barriers.

G. K. GILBERT'S PHILOSOPHY OF RESEARCH

In investigating the history of Lake Bonneville and other Quaternary water bodies of the Great Basin, the writer and his assistants have had constant occasion to distinguish from all others the element of topography having a littoral origin and have become familiar with the criteria of discrimination.

Gilbert (1885, p. 75) thus introduced his early work on the origin and evolution of coastal systems, work that led to many visits to the shorelines of Lake Bonneville, the Great Lakes, and the Atlantic and Pacific coasts.

Gilbert had very little literature upon which to base his studies. It is therefore useful to note his philosophy of research. No better is this stated than in his paper "The Inculcation of Scientific Method by Example" (1886, p. 25):

It is the province of research to discover the antecedents of phenomena. This is done by the aid of hypothesis. A phenomenon having been observed, or a group of phenomena having been established by empiric classification, the investigator invents an hypothesis in explanation. He then devises it and applies a test of the validity of the hypothesis. If it does not stand the test he discards it and invents a new one. If it survives the test, he proceeds at once to devise a second test. And he thus continues until he finds an hypothesis that remains unscathed after all the tests his imagination can suggest. This, however, is not his universal course. For he is not restricted to the employment of one hypothesis at a time. There is indeed an advantage in entertaining several at once, for then it is possible to discover their mutual antagonisms and inconsistencies, and to devise crucial tests—tests which will necessarily debar some of the hypotheses from further consideration. The process of testing is then a process of elimination, at least until all but one of the hypotheses have been disproved. . . . Evidentally, if the investigator is to succeed in the discovery of veritable explanations of phenomena, he must be fertile in the invention of hypotheses and ingenious in the application of tests.

GILBERT'S METHODS IN THE STUDY OF BARRIERS

G. K. Gilbert's initial ideas and hypotheses regarding the origin of barriers came to him in his studies of the ancient shorelines of Lake Bonneville; however, he recognized the importance of the modern analog, and in the succeeding years, he observed shorelines in New England, the Great Lakes, and central California (see Figs. 1, 2, 3). In these places he observed modern barriers and lagoons that were analogous to certain shoreline features of Pleistocene Lake Bonneville.

While studying the shorelines of Lake Bonneville, Gilbert quickly moved from observation of morphology to hypothesis regarding coastal process. In testing hypotheses, Gilbert was greatly aided by gullies on whose walls he was able to note details of the internal structure of the barriers (1885, Pl. IV; 1890, Figs. 12, 13). Gilbert recognized the importance of understanding internal structure in determining the origin of the landforms. Gilbert (1885,

p. 77) was aware that this gave him an advantage over the coastal engineer.

The geologic student has, too, some facilities for study which the engineer lacks, for he is frequently enabled to investigate the anatomy of shore structures by means of natural cross-sections, while the engineers are restricted to an examination of their superficial forms.

This is the essence of Gilbert's contribution to the study of the evolution of coastal barrier systems. He recognized that an understanding of processes, physiography, and internal structure was necessary to prove a hypothesis on the origin of barriers.

LITTORAL TRANSPORT

The most fundamental idea conceived by Gilbert was the importance of the agent littoral transportation (littoral transport) in the formation of spits and barriers and that the driving mechanism of littoral transport was the joint action of waves and currents. He noted that the source and type of sediment or rock undergoing erosion will form the primary control on the size of particles transported. As littoral transport progressed, separation or division of sediments into different sizes would tend to occur. Gilbert noted the concept that doubling the velocity of a current more than doubles the amount of sediment that can be carried and the size of particles that can be moved. In his discussion of waves and currents, he noted:

An incoming oblique wave transfers the particles in the direction of the wave while the background transfer, by means of the undertow, is sensibly normal to the shore. . . . The littoral current thus tends in a direction harmonious with the movement of the waves. [1885, p. 85, 86]

The material transported, called shore drift (littoral drift) by Gilbert, was declared to be the dominant mechanism for the formation of barriers. An important concept noted in his observations of lake and ocean shorelines was that the littoral transport followed the line of breakers instead of the water margin or shoreline. Thus a ridge could be extended into a water body.

It will be convenient to speak of this ridge as a barrier. . . . The barrier is the functional equivalent of the beach. It is the road along which shore drift [littoral drift] travels, and it is itself composed of shore drift. . . . Behind the

Figure 1. Bar (barrier) joining Empire and Sleeping-Bear Bluffs, Lake Michigan (Gilbert, 1885). Lake Michigan on left, barrier in center, lagoon to the right.

Figure 2. Bolinas Lagoon, California, photographed by Gilbert (compare Bergquist, 1979). Pacific Ocean on left, barrier-spit center, eroding cliffs center-background, lagoon to the right. Photograph no. 3050, U.S. Geological Survey, 2 hours before low water, March 8, 1907.

barrier and the land a strip of water is enclosed, constituting a lagoon. This is frequently converted into a marsh by the accumulation of silt and vegetable matter and eventually becomes completely filled so as to bridge over the interval between the land and the barrier. . . . Beach and barrier are absolutely dependent on shore drift [littoral drift] for their existence. [1885, p. 87, 88]

For his first modern analog, Gilbert went to the Great Lakes where he could observe actual headland erosion, littoral transport, and the deposition of a spit which eventually evolved into a barrier (Figs. 1, 3). He recognized the dominant driving force was that of waves and currents. Gilbert (1917, p. 69) noted the great importance of littoral transport along the Pacific coast north of the Golden Gate.

For seven miles north of the Golden Gate the coast presents a series of cliffs, and it is evident that these are sapped and eaten back by the beating of the waves. The waste from that erosion is drifted by shifting currents in two directions. Part of it reaches Bolinas Lagoon (Figure 2), where it is built into a bar across the entrance, and another part goes past Point Bonita to the entrance of the Golden Gate. Thirteen miles south of the gate, at San Pedro point, begins another line of cliffs from which the waste has swept in two directions. The larger part goes northward. . . . A portion of its sand escapes from the sea and is blown inward, where it travels in dunes; another portion finds it way, either directly ir indirectly, to the Golden Gate bar. . . . The mode of progression of the cliff-born sand involves a cooperation of waves and currents. Each wave disturbs the sand of the beach, momentarily lifting a portion, and while the grains are free from the bottom they drift with the current.

Gilbert concluded that the products of littoral transportation (littoral transport) were mainly derived from land erosion. He noted that probably the only difference between ocean and lake systems would be that of tides or lack of tides.

Probably the most important specific location upon which Gilbert based his hypotheses was the "wave built barrier" across

the Rush and Tooele Valleys near Stockton, Utah. Here a barrier had evolved by the littoral transport of materials eroding from an adjacent shoreline and deposition of these materials on a spit which had been alternately connected to the headland to the east of the valleys (Figs. 3, 4). The exhumed barrier-lagoon system could be observed in great detail. H. A. Wheeler mapped the system and drew a profile section across the barrier showing that it had evolved at a lower lake level and migrated lakeward to its peak development at the highest level of Lake Bonneville (Fig. 3). A perspective sketch (Fig. 4) easily enables the reader to envisage what Gilbert had in mind. In viewing Figure 4, one should visualize the "bowl" filled with water, a lake to the left, a barrier in the center, and a lagoon to the right.

In an attempt to understand the shoreline systems of Lake Bonneville, Gilbert educated himself to the state of wave science of his day. He noted that wind waves were probably more efficient and more important than swells in terms of producing water movement that would transport particles (Fig. 5).

[As the wave breaks] a particle of water at or near the surface, as each undulation passes, describes an orbit in a vertical plane, but does not return to the starting point. While on the crest of the wave it moves forward, and while in the trough, it moves less rapidly backwards, so that its residual advance is the access of one movement over the other.[Gilbert, 1885, p. 80]

Gilbert freely acknowledged that his concepts on wave processes were based on the works of many engineers and scholars of the time.

In his work along the Pacific coast in central California, Gilbert (1917) probably recognized that the spit at Bolinas Lagoon is a modern marine analog to the barrier at Stockton, Utah, along the shoreline of Lake Bonneville. Adjacent headlands were undergoing erosion, littoral transport was carrying littoral drift in a spitlike fashion across the entrance of an embayment, and finally

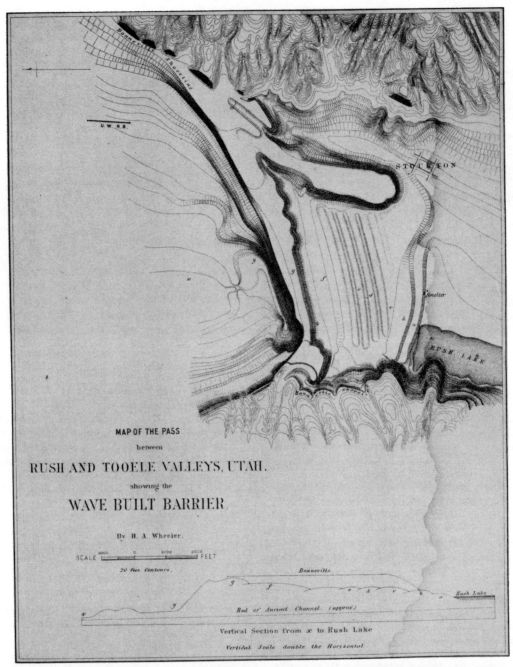

Figure 3. Topographic map of the exhumed barrier lagoon system in the pass between Rush and Tooele valleys, Utah, showing the wave-built barrier, as surveyed by H. A. Wheeler (Gilbert, 1890).

nearly sealing it off to form a barrier-spit and lagoon. In the fringes of San Francisco Bay, he noted that marshes were building upward and had the potential of filling portions of the bay. East of Larkspur, in Marin County, California, Gilbert photographed a drowned topography surrounded and being encroached upon by a salt-water marsh (Fig. 6). He noted that drowned or emergent topography was based on rise and fall of the land. In view of these observations, he further concluded that a potential course existed for complete landward migration of a beach to the point where wave-cut cliffs totally bypassed and destroyed lagoons.

Gilbert noted that storm processes were dominant in coastal evolution. This was both in terms of their ability to move larger particles and in the total volume of sediment that might be moved. He was also aware of the rapid reconstruction of the beach and berm after a storm had passed.

In much of Gilbert's coastal terminology, it is difficult to determine whether or not he was using terms and concepts from the vernacular or developing new ones. But the concept and terminology of "shore drift" (littoral drift) were clearly developed by Gilbert. Gilbert used the words "bar" (barrier) and "spit" in 1885. The essence of his hypothesis for the origin of barriers was stated in his 1885 publication in which he discussed "the spit": "When a

Figure 4. Perspective diagram of the barrier near Stockton, Utah, as sketched by Gilbert (1885). Note the Lake Bonneville basin to the left, the linear barrier, and the isolated lagoon in the valley to the right. Sediment was eroded from the cliffs in the far left of the picture, and transported by littoral transport, and deposited on a spit building across the embayment. When this spit reached the headland in the foreground, it formed a barrier.

coastline followed by a littoral current turns abruptly landward, as at the entrance of a bay, the current does not turn with it, but holds it course and passes from shallow to deeper water" (1885, p. 91). Gilbert noted the problem of potential refraction around the spit thus forming "hooks" (recurved spits) but pointed out that if the supply of shore drift (littoral drift) did not cease, the particles might continually be carried forward and dropped into the deeper water body forming a barrier, which he compared to the construction of a railroad embankment in which railroad cars carry

the material forward and dump it. "The form is often more perfect than the railway engineer ever accomplishes" (1885, p. 92). Going on, speaking of the bar (barrier) Gilbert added:

If the current determining the formation of a spit again touches the shore, the construction of the embayment is continued until it spans the entire interval. So long as one end remains free the vernacular of the coast calls it a *spit;* when it is completed it becomes a *bar* [barrier]. [1885, p. 92]

By 1890 Gilbert was aware that transgressions could occur across a barrier with the barrier and lagoon migrating landward. This could be formed by a rise in water level or land subsidence. The result would be a wave-cut terrace on a transgressive barrier. Equally, should shore drift (littoral drift) provide sufficient material, the barrier would tend to build lakeward or seaward, thus resulting in the geological phenomenon known as beach accretion or a resultant regressive barrier. Gilbert referred to such features as a wave-built terrace (regressive barrier)

The quantity of shore drift moved depends upon the magnitude of the waves; but the speed of transit depends upon the velocity of the current, and

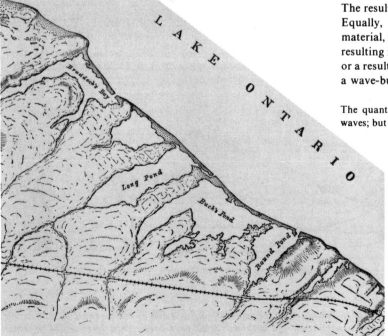

Figure 5. Map of spits and barriers along the southern shore of lake Ontario (Gilbert, 1885). Gilbert regarded these as modern analogs for shoreline features of Pleistocene Lake Bonneville (compare Fig. 4).

Figure 6. Tidal marsh east of Larkspur, looking east. A land surface of varied topography is drowned by marsh deposits, all hill tops being left as islands. Marin County, California, 1906, photograph by G. K. Gilbert, no. 2985, U.S. Geological Survey.

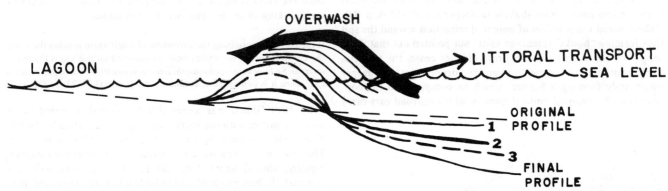

Figure 7. Schematic drawing showing de Beaumont's hypothesis for the evolution of coastal barriers by emergence from a submarine bar (after Otvos, 1979). Littoral transport plays a minor role in construction of the bar but a major role in the continuing growth of the bar after it emerges.

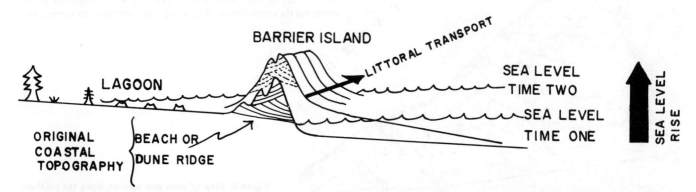

Figure 8. Schematic section showing Hoyt (1967) and McGee (1890) hypothesis of the formation of barrier islands by the drowning of coast-parallel geomorphic features by relative sea-level rise. Littoral transport plays no role in the origin of the barrier but has a major role in the growth of the barrier once initiated.

wherever that velocity diminishes, the accession of shore drift [littoral drift] must exceed the transmission, causing accumulation to take place. This accumulation occurs, not at the end of the beach, but on its face, carrying its entire profile lakeward [seaward] and producing by the expansion of its crest a tract of new-made land. [1885, p. 97]

Gilbert also noted the effect of undertow in carrying away clay in the littoral transport of sands (1885, p. 99). Here Gilbert in effect explained the concept of negative skewness of beach sands. He further noted that sands could be blown across barriers into lagoons, thus encroaching or infilling lagoons or bays and leading to their eventual destruction. In an illustration of the shoreline of Lake Superior near Marquette, Gilbert (1885, Pl. XV) illustrated a tombolo and noted that its origin was via the process of littoral transport. Apparently he did not recognize the offshore "breakwater" effect of the outliers typical of tombolos.

THE ORIGIN OF BARRIER ISLANDS

A great volume of literature has been generated since Gilbert's day regarding the origin of coastal barriers. Most researchers now seem to accept "multiple causality of barrier islands" (Schwartz, 1971; Fisher, 1979). Davis (1979) edited a book of articles presenting modern thoughts on coastal sedimentary environments in total, and Leatherman (1979) edited a book of articles on the latest concepts of barrier island structures and evolution.

The idea of de Beaumont—shoals rising above sea level—seems to be held invalid by many researchers, or regarded as merely a minor factor in the origin of barriers. However, it is interesting to note that R.A.D. Davis, Jr. (1979, oral commun.) at the University of South Florida has recently called attention to the fact that barrier islands have emerged from shoals along the southwestern coast of the Florida peninsula in recent historic time. Yet even here it is possible that littoral transport by ebb-tide processes might in fact be feeding the shoals that are "rising out of the sea." Along the eastern shoreline of the Gulf of Mexico, Otvos (1970, 1979) has identified a barrier island chain that has apparently evolved according to the concepts of de Beaumont (Fig. 7).

Hoyt's hypothesis of a drowned topography being realigned into barriers also appears to be valid, as evidenced in his article of 1967, and strongly supported by Oertel (1979). Figure 8 summarizes the concepts of Hoyt (1967) and McGee (1890). Littoral transport processes are important in the continuing development and growth of coastal barriers in both the de Beaumont and Hoyt-McGee models.

Along the Atlantic, Gulf Coast, Pacific, and Great Lakes shorelines of the United States, sediment moves primarily in the littoral transport stream. This leads to the conclusion that Gilbert's hypothesis on the origin of barrier islands is a most viable one. If this statement is limited to the slightly different concept that littoral transport is *the* major process of continuing buildup and modification of barriers, rather than of initiation, then some of the ongoing argument may be put to rest. One cannot deny the existence of littoral transport and its overwhelming importance along the coastlines of the United States, particularly in the barrier island chains of the Atlantic Ocean and Gulf of Mexico.

Barrier islands can both transgress or regress, dependent upon tectonic, eustatic, sedimentologic, and climatologic factors. Recent studies of the barriers of the Atlantic and Gulf coasts suggest that, to a large extent, they are transgressive (Kraft, 1971; Kraft and others, 1979; Wilkinson and Basse, 1978). If so, then some of these barriers probably formed elsewhere and migrated to their present position (Fig. 9). Regressive barriers, such as Galveston Island (Fig. 10), can of course be studied from the point of their initiation. Certainly the beach accretion ridges of Galveston Island indicate the importance of littoral transport in moving sediment into place. However, even here, enclosure of an embayment by spit progradation may have initiated regression by the beach accretion plain (Fig. 10).

It is possible that some of the transgressive barriers formed in early Holocene time on the outer continental shelf and migrated to their present position. If so, then it becomes an exercise in futility to investigate the origin of transgressive barrier inlands by studying these islands in their present positions. From the point of view of regressive barrier islands, it should be possible, with careful three-dimensional studies of the sedimentary environmental units involved, to elucidate the origin of such barrier islands in individual cases. In either case, it appears unlikely that a single hypothesis will suffice as a universal explanation for the origin of barrier islands.

We live in an interval of geologic time known as the Holocene Epoch. The most overwhelming event of this time interval is that of the rise of sea relative to land with the waning of the last major glacial ice sheets. With this waning occurred many complex tectonic movements related to the margin of the ice sheet, the effect of water loading on the continental shelves, and other relative vertical land movements. Although researchers have varied opinions, it appears probable that sea level either reached its present position from 4,000 to 6,000 yr B.P. or sharply slowed down in its rate of rise relative to land. At this particular time, most of the barrier islands of the Atlantic and Gulf coast probably began to form.

CONCLUSIONS

Where does G. K. Gilbert stand in the investigation of barriers and spits? He was aware of nearly all of the major coastal processes that are important in the formation and migration of barrier islands.

Gilbert originated the hypothesis of littoral transport, a process that accounts for most of the world's barrier islands and affects most of the others. Nevertheless, the development of many barriers seems to be consistent with the early concepts of de Beaumont and McGee. Major advances beyond Gilbert's work have been based on study of the geometry of coastal facies and the nature of coastal processes.

In essence, one could say "G. K. Gilbert was right." However, G. K. Gilbert would probably have objected to this view. Gilbert never stated that he was building a comprehensive theory on the origin of barrier islands. He emphasized instead that he was merely attempting to explain some features along the shorelines of Pleistocene Lake Bonneville by the use of modern coastal analogs.

ACKNOWLEDGMENTS

I thank Ellis Yochelson, Clifford Nelson, Jack Pierce, and Brian Atwater for their helpful editorial comment. I also wish to thank Ellis Yochelson and Joel Bergquist for their help in locating Gilbert's photographs used herein.

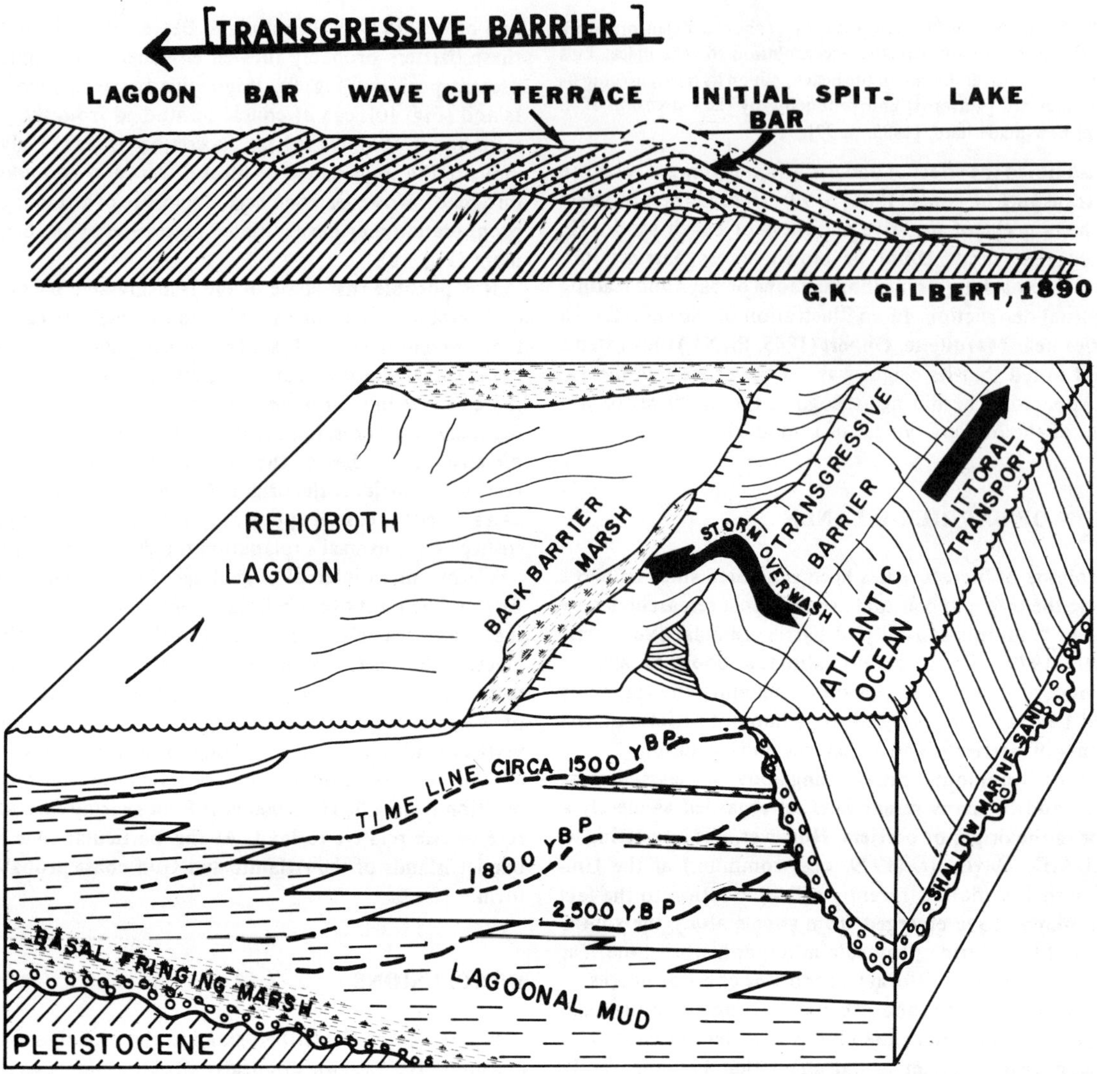

Figure 9. Perspective diagram of a transgressive coastal barrier along the Atlantic coast of Delaware. This barrier originated farther seaward on the Atlantic Continental Shelf and has migrated to its present position because of relative sea-level rise (Kraft and others, 1979; Kraft and John, 1979). Time-depositional lines show the surface of the barrier and associated coastal environments as the shoreline moved landward and upward with time. Storm overwash processes and littoral transport play a dominant role in the formation of this type of barrier. Essentially, Gilbert's shore-drift (littoral transport) hypothesis is in action in the continuing migration of this barrier. Gilbert's cross section of a barrier retreating landward with the construction of a wave-cut terrace (1890, Fig. 12) is shown at top for comparison. Note that Gilbert shows initial foresets toward lagoon and lake as the spit forms. With the initiation of transgression, the wave-cut terrace is cut and dominant; deposition is foreset into the lagoon as the barrier migrated landward.

REFERENCES CITED

Bergquist, J. R., 1979, Bolinas Lagoon, Marin County, California: California Geology, v. 32, no. 10, p. 211–216.

Bernard, H. A., and LeBlanc, R. J., 1965, Resume of the Quaternary geology of the northwestern Gulf of Mexico province, *in* Wright, H. E., and Frey, D. G., eds., The Quaternary of the United States: Princeton, Princeton University Press, p. 137–185.

Davis, R. A. D. Jr., 1979, Coastal sedimentary environments: New York, Springer-Verlag, 420 p.

De Beaumont, E., 1845, Septiéme leçon: lecons de geologie pratique: Levéés de sable et de galet, Paris, P. Bertrand, v. 1, p. 221–253.

Fisher, J. J., 1979, Barrier islands: Pre-publication manuscript *in* Schwartz, M. L., ed., Encyclopedia of beaches and coastal environments, Volume XV: Stroudsburg, Dowden, Hutchinson and Ross, Inc.

Gilbert, G. K., 1885, The topographic features of lake shores: p. 69–123 *in* Fifth Annual Report of the U.S. Geological Survey: 1883–84: U.S. Government Printing Office, 469 p.

——— 1886, The inculation of scientific method by example: American Journal of Science, ser. 3, v. 31, p. 284–299.

——— 1890, Lake Bonneville: U.S. Geological Survey Monograph 1, 438 p.

——— 1900, Rhythms and geologic time: Science, v. 11, p. 1011–1012.

——— 1917, Hydraulic mining debris in the Sierra Nevada: U.S. Geological Survey Professional Paper 105, 154 p.

Hoyt, J. H., 1967, Barrier island formation: Geological Society of America Bulletin, v. 78, p. 1125–1136.

Johnson, D. W., 1919, Shore processes and shoreline development: New York, John Wiley & Sons, 584 p.

Kraft, J. C., 1971, Sedimentary facies patterns and geologic history of a Holocene marine transgression: Geological Society of America Bulletin, v. 82, p. 2131–2158.

Kraft, J. C., and John, C. J., 1979, Lateral and vertical facies relation-

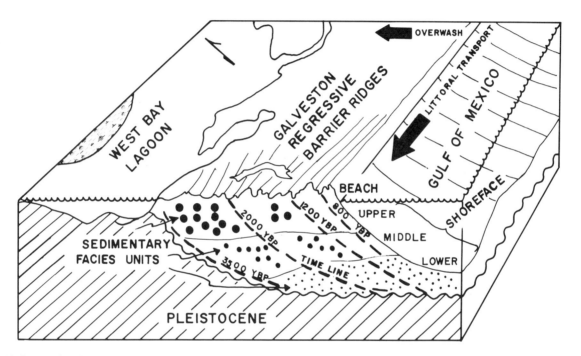

Figure 10. Perspective diagram showing the internal geometry of the Galveston regressive barrier (after Bernard and LeBlanc, 1965; LeBlanc and Hodgson, 1959). The Galveston barrier may have originated by drowning of a coast-parallel feature. However, from its immediate initiation, the supply of littoral drift from adjacent eroding headlands and deltas was the dominant process in the formation of this regressive barrier island. A schematic section of a "wave-built terrace" (beach accretion plain) by Gilbert (1890, Fig. 13) is shown for comparison. Gilbert's hypothesis fits the origin of this regressive barrier island very closely.

ships of a transgressive barrier: American Association of Petroleum Geologists Bulletin, v. 63, p. 2145–2163.

Kraft, J. C., and others, 1979, Processes and morphologic evolution of an estuarine and coastal barrier system, in Leatherman, S. P., ed., Barrier islands from the Gulf of St. Lawrence to the Gulf of Mexico: New York, Academic Press, p. 149–183.

Leatherman, S. P., 1979, Barrier islands from the Gulf of St. Lawrence to the Gulf of Mexico: New York, Academic Press, 325 p.

Leblanc, R. J., and Hodgson, W. D., 1959, Origin and development of the Texas shoreline: Gulf Coast Association Geological Society Transactions, v. 9, p. 197–220.

McGee, W. J., 1890, Encroachment of the sea: The Forum, v. 9, p. 437–449.

Mitchell, H., 1869, On the reclamation of tide-lands and its relation to navigation: Appendix No. 5 to the Report of United States Coast Survey for 1869, Washington, 1872, p. 75–104.

Oertel, G. F., 1979, Barrier island development during the Holocene recession, southeastern United States, in Leatherman, S. P., ed., Barrier islands from the Gulf of St. Lawrence to the Gulf of Mexico, New York, Academic Press, p. 273–290.

Otvos, E. G., Jr., 1970, Development and migration of barrier islands, northern Gulf of Mexico: Geological Society of America Bulletin, v. 81, p. 241–246.

——1979, Barrier island evolution and history of migration, north central Gulf coast, in Leatherman, S. P., ed., Barrier islands from the Gulf of St. Lawrence to the Gulf of Mexico: New York, Academic Press, p. 291–319.

Pierce, J. W., and Colquhoun, D. J., 1970, Holocene evolution of a portion of the North Carolina coast: Geological Society of America Bulletin, v. 81, p. 3697–3714.

Schwartz, M. L., 1971. The multiple causality of barrier islands: Journal of Geology, v. 79, p. 91–94.

Wilkinson, B. H., and Basse, R. A., 1978, Late Holocene history of the central Texas Coast from Galveston Island to Pass Cavallo: Geological Society of America Bulletin, v. 89, p. 1592–1600.

MANUSCRIPT RECEIVED BY THE SOCIETY JANUARY 7, 1980
MANUSCRIPT ACCEPTED MAY 20, 1980

Geological Society of America
Special Paper 183
1980

G. K. Gilbert and the great iceberg-calving glaciers of Alaska

MARK F. MEIER
AUSTIN POST
U.S. Geological Survey, Tacoma, Washington 98402

ABSTRACT

G. K. Gilbert, as a member of the Harriman Alaska Expedition of 1899, studied and described nearly 40 glaciers, many of which reached the sea and produced icebergs. Gilbert's maps and photographs from marked locations are still being used to record glacier fluctuations, as at Columbia Glacier. Noting that some termini were stable or advancing but that others were retreating rapidly, he suggested that a general change in climate, perhaps related to a change in ocean temperature, might cause such local differences in behavior. This conclusion was remarkably prescient, but it is now known that terminus stability is also involved. Gilbert's discussions of the processes of glacier flow adjustment to an uneven bed, glacial erosion (including erosion below sea level), and variations in the rate of iceberg calving are remarkably modern and relate to one of the most important problems in glaciology today—the role of a water layer in coupling a glacier to its bed.

GLACIOLOGY IN THE NINETEENTH CENTURY

To the great naturalists of the eighteenth and nineteenth centuries, the mighty glaciers of the Alps were a source of fascination. How could one explain the paradox of rigid, brittle ice flowing down a valley? What did the ample evidence of previous glacier advances and retreats tell us about climate?

The study of glaciers did not begin in the United States until the late nineteenth century, particularly when naturalists and geologists such as John Muir, G. F. Wright, I. C. Russell, H. F. Reid, and G. K. Gilbert visited the huge coastal glaciers of Alaska. Usually traveling by boat, they were especially fascinated by those glaciers having awesome terminal ice cliffs which discharged (calved) icebergs into the sea (Figs. 1, 2).

These spectacular glaciers posed some fascinating new scientific problems. In addition to his important research on problems of Pleistocene glaciation, Gilbert was concerned with questions such as: How could a glacier erode rock far below sea level even though being partially lifted off that rock by the buoyant force of the sea? What controlled the variations of these glaciers, and why were some glaciers in very rapid retreat and others essentially stable or slowly advancing? These questions are interrelated, and have become an important focus on U.S. Geological Survey glaciological research at the present time.

THE HARRIMAN ALASKA EXPEDITION

G. K. Gilbert's reputation in science was so firmly established in 1899 that "when a great excursion to Alaska was planned. . .and a fine ocean-going vessel chartered for the voyage from Seattle to Alaska and return by the host of the party, Mr. Edward H. Harriman, it was natural if not inevitable that Gilbert should be included among its 25 scientific members, inasmuch as they were chosen as leading representatives in various lines of research" (Davis, 1927, p. 229).

Gilbert realized the importance of documenting the fluctuations of the large and varied set of Alaskan calving glaciers. He described nearly 40 individual glaciers, many of which reached tidewater. He visited as many as possible of those that earlier observers had described or mapped to measure their changes. At many of those not previously studied, he compiled maps, photographs, and descriptions so their changes could be measured by later observers. So numerous were Alaskan glaciers, and so little had they been explored, that this expedition discovered and named several major ice streams.

ASYNCHRONOUS FLUCTUATIONS OF CALVING GLACIERS

It is fortunate that Gilbert and other early investigators made such careful observations. Since the time of Gilbert's studies, tremendous changes have occurred in several tidewater glaciers in Alaska: Muir, Guyot, Northwestern, and McCarty Glaciers have undergone rapid and long retreats; others, such as the Taku, Lituya, Hubbard, and Harvard, have made slow but continuous advances. The asynchronous behavior of calving glaciers is well illustrated by the comparison of the Muir and Columbia Glaciers. Muir, the glacier that received special attention by Reid (1892, 1896) and considerable study by Gilbert (Fig. 3), was retreating at a rate "perhaps without parallel in the records of glacial changes within the historic period" (Gilbert, 1903, p. 17). Today it has almost disappeared, opening up the huge fiord system of Muir Inlet in Glacier Bay (Fig. 4). On the other hand, Columbia Glacier today appears much the same as it did to G. K. Gilbert (Figs. 5, 6).

Gilbert recognized the dilemma of these asynchronous variations: "The most conspicuous fact brought out by the comparison of local histories is that they are dissimilar. . . . It is natural to look

Figure 1. Columbia Glacier, named and first studied by G. K. Gilbert. U.S. Geological Survey photograph, August 22, 1979, by Austin Post.

Figure 2. The 90-m-high calving face of Columbia Glacier. U.S. Geological Survey aerial photograph by L. R. Mayo, October 8, 1975.

for a systematic geographic arrangement of the diverse histories; but such an arrangement is not apparent" (Gilbert, 1903, p. 104-105). The two theories in vogue at the time were rhythmic fluctuations of the néve reservoir, with the period different from one glacier to the next; or varying "lags" (response times), with the lag different from one glacier to the next. To Gilbert, neither theory explained the observations in Alaska. In an ingenious discussion, Gilbert surmised that a general climate change, perhaps in the form of a general change in the ocean temperature, could "result in local glacier variations which are not only unequal but opposite." (Gilbert, 1903, p. 109). Only in the last several decades have meteorologists and climatologists begun to focus attention on the critical effect of ocean temperatures on meteorologic processes on land, and on the complex and asynchronous responses of glacier mass balances to climate change.

Thus, Gilbert was correct in his general explanation of many Alaskan variations. However, the contrast in behavior between Muir and Columbia Glaciers, and the drastic rate of change of the Muir, cannot be explained by climate alone. It has since been found that the rapidly retreating (unstable) tidal glaciers have all uncovered deep fiords. Advancing and stationary (stable) glaciers terminate either on moraine shoals or at the heads of fiords where bedrock rises above sea level. Stability, or instability leading to irreversible rapid retreat, depends primarily on the water depth at the terminus (Post, 1975).

THE COLUMBIA GLACIER

One of the most noteworthy side trips taken by Gilbert during the Harriman Alaska Expedition was "one in Prince William Sound, where, with Coville and Palache, he explored and mapped the most stupendous glacier visited by the expedition, a glacier having a sea-wall frontage of four miles. This he named after the geologist and explorer I. C. Russell, but later, finding Russell's name preoccupied by a glacier in the Copper River region, this one was rechristened the Columbia" (Merriam, 1919, p. 394-395). Columbia Glacier (Fig. 7) is presently unique as the only Alaskan glacier that still terminates near its maximum neoglacial ("Little Ice Age") position. All others have made drastic retreats. Should Columbia Glacier begin a drastic retreat, the volume of icebergs discharged would increase, possibly endangering oil tankers enroute to and from nearby Valdez, southern terminal of the Trans-Alaska Pipeline. Thus, the future changes in this glacier are of economic as well as scientific interest. As a result, the U.S. Geological Survey has made intensive studies of this glacier, and it is indeed fortunate that the data base was begun so early by G. K. Gilbert.

Moving from point to point around Columbia Bay by boat, Gilbert went ashore at selected localities and established photographic stations from which present and future photography could be obtained for detailed comparison. A map of the lower part of the glacier was compiled (Fig. 5), and observations of many features were recorded. Gilbert, unlike many others, did not hesitate to exert himself in difficult ascents to high vantage points on nearby mountains from which sweeping views of the glacier could be had. This strenuous habit revealed many important features which would have otherwise been missed.

Many of Gilbert's photographic and mapping stations at Columbia Glacier have since been reoccupied by later researchers, including Grant and Higgins (1913), Tarr and Martin (1914), and Field (1932, 1974). By comparing conditions with Gilbert's earlier

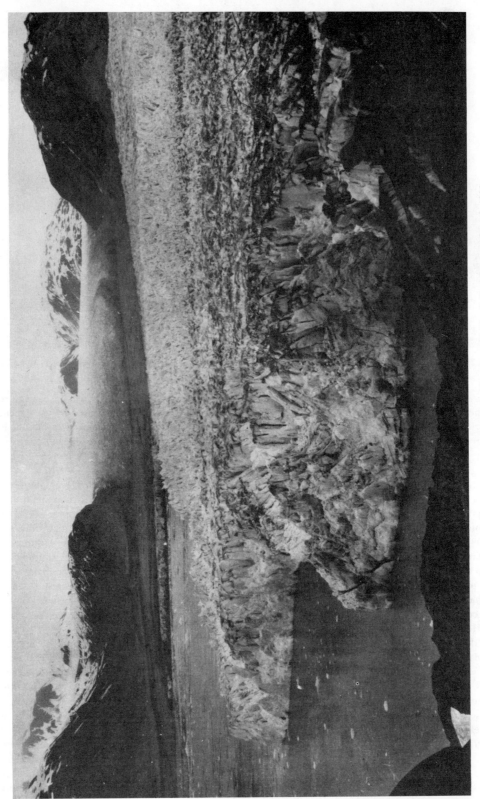

Figure 3. The calving terminus of Muir Glacier, as photographed by G. K. Gilbert on June 9, 1899. The terminus was about 4 km wide at this time. Photo location shown on Figure 4.

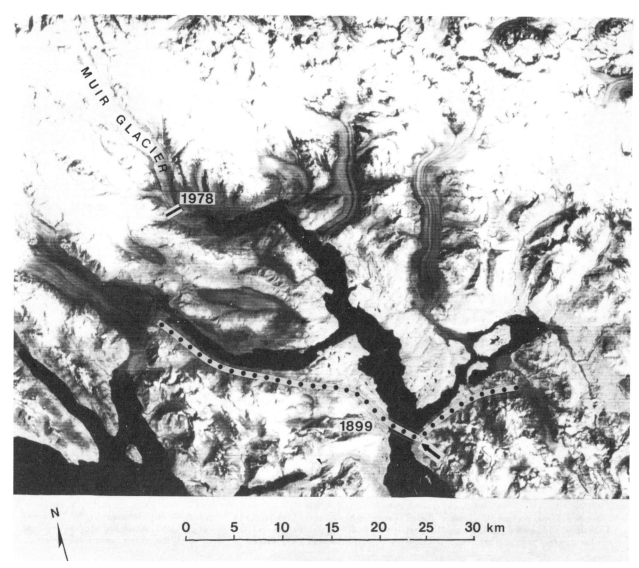

Figure 4. Satellite image of Muir Glacier on July 30, 1978 (Landsat (ERTS) image 30147-19375-7). Dotted line shows terminus as it appeared in Gilbert's time; virtually all the land and water area to the north was then covered with ice. Arrow indicates the location and direction of the view in Figure 3.

records, these observers were able to document advances that culminated in 1910, 1921, and 1935. The authors were fortunate enough to witness a smaller, although impressive, advance of the glacier in 1974 on Heather Island. Here the ice was pushing up a wall of moraine; by remarkable coincidence the crest of this debris was placed directly over Gilbert's station "E." From this point the glacier has since withdrawn; perhaps this was the last advance of Columbia Glacier on Heather Island, as its present terminal stability is precarious.

Although the terminus of Columbia Glacier has changed only slightly since 1899, Gilbert's data reveal that an interesting and important change has occurred farther upglacier. His map and descriptions show that medial-moraine positions were different from those of today; the East Branch of the glacier contributed far more ice then than now. The main glacier is supplied from icefields at much higher altitudes, suggesting that a rise in snowline has occurred, which results in reduced snow accumulation at lower

levels but with less relative loss at higher altitudes, so that the discharge of the East Branch would be reduced relative to that of the main glacier.

GILBERT'S CONTRIBUTIONS TO THE PHYSICS OF GLACIERS

As a result of his observations on Columbia and other calving glaciers, Gilbert recognized and addressed the following problems which remain, or have reappeared, prominent in glaciological thought today. Why is the surface of a glacier more even than the bed? How does glacial erosion occur, especially the excavation of fiords far below sea level? What controls iceberg calving? All of these questions relate to how a glacier is coupled to a bed through its water system.

Gilbert considered the contrast between the large-scale smooth-

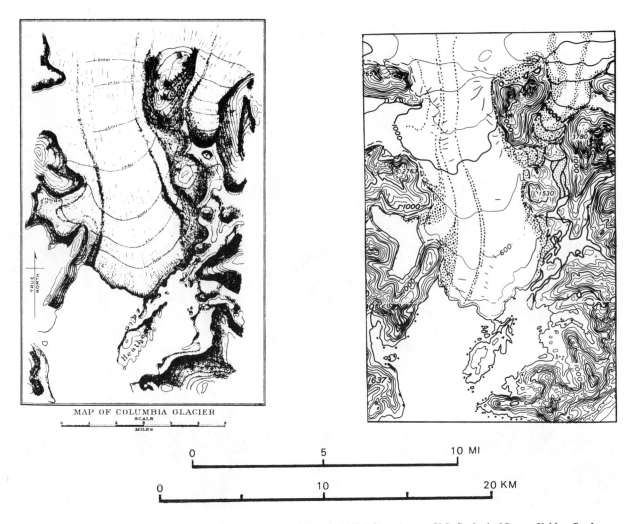

Figure 5. Lower portion of Columbia Glacier, as mapped by Gilbert in 1899 (left), and recent U.S. Geological Survey Valdez, Cordova, Anchorage, and Seward 1:250,000 quadrangles (right). The only major changes are (1) the glacier-dammed lake northwest of the terminus has become free of ice, (2) the medial moraine has become separated from the east margin, and (3) the glacier has generally thinned.

ness of a glacier surface and the roughness and unevenness of the bed. He stated, "Variations of direction [of flow] alone are not sufficient to explain the coexistence of an approximately even glacier surface with a very uneven bed. The condition of continuity cannot be satisfied without variations of velocity also within the mass" (Gilbert, 1903, p. 197). This statement, in language that seems very modern, is also remarkably prophetic: perhaps the most difficult task in modeling the flow of Columbia Glacier at the present time is the assembling of a suite of data that satisfies the continuity condition at every point. The irregularities of the bed and the spatial variation in basal sliding affect the velocity components throughout the glacier.

Processes of glacial erosion were poorly understood in Gilbert's time and remain so today. Some of Gilbert's arguments, although couched in purely descriptive or qualitative terms, parallel modern theory. Considering ice as a viscous liquid, Gilbert (1903, p. 204) suggested that erosion tends to produce a bed comprised of longitudinal concave curves of large radii, the radius being a function of the basal ice velocity. Nye and Martin (1968) came to a somewhat similar conclusion by considering ice as a perfectly plastic solid; they also suggested that erosion tends to produce a

bed comprised of arcs with large radii, but that the radii are primarily functions of ice thickness. Since thickness and velocity are related, these two theories predict similar effects.

In a section entitled "Pressure and erosive power of tidal glaciers," Gilbert (1903, p. 201–218) addressed a subject critical to his geomorphical studies of fiords: Inferred erosion far below sea level. These studies brought him to consideration of a water film between ice and rock, as he realized that glacial erosion below sea level depends on whether a water film is or is not connected hydraulically with the seawater which exerts hydrostatic (buoyant) pressure back under the glacier. He suggested that a film of infinitesimal or capillary thickness would not prevent the rock bed from supporting the whole weight of the glacier, and thus that erosion could occur. On the other hand, a thicker film or a conduit with flowing water would transmit the hydrostatic pressure of the seawater, causing it to contribute to the support of the glacier, reducing or eliminating erosion. He then conceived of a tidal-glacier bed as containing a mixture of areas of no film, of a film of capillary thickness, and of a thicker stratum of water. If we replace the word "capillary" with "small," the concept is in agreement with today's theories.

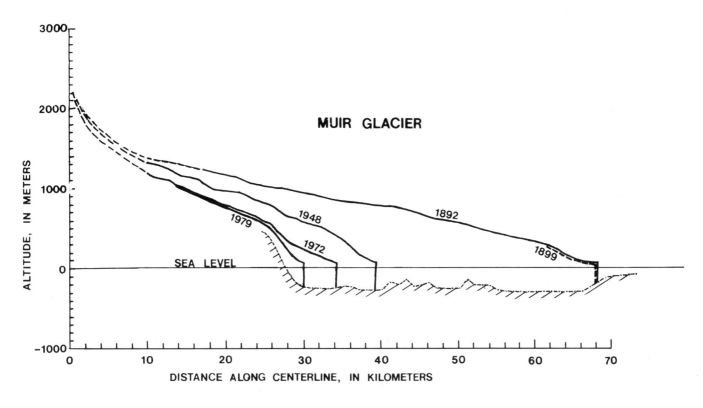

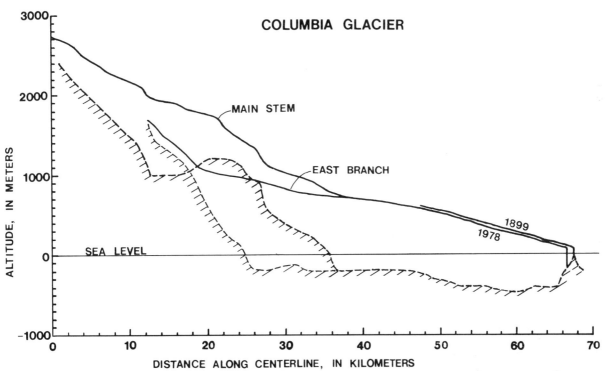

Figure 6. Longitudinal profiles of Muir Glacier (upper) and Columbia Glacier (lower). Profile of Muir Glacier in 1892 from Reid (1896); later profile from U.S. Geological Survey topographic maps. Surface profile of Columbia Glacier from 1978 surveys; bed profile from 1974 to 1979 monopulse radar soundings, and computer solution using the equation of continuity.

Figure 7. Columbia Glacier; photograph taken in 1899 by G. K. Gilbert from a point about 8 km to the south.

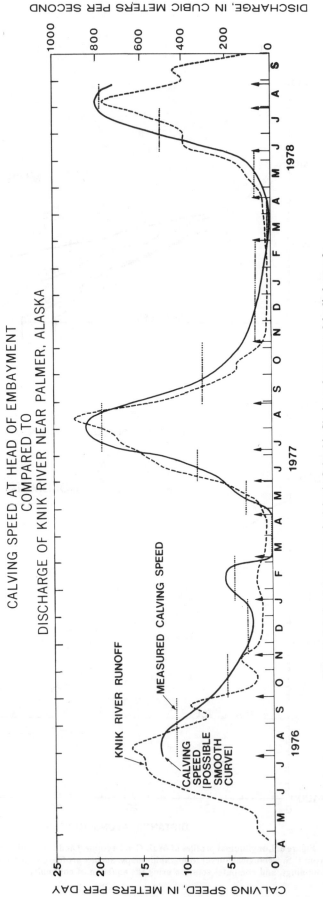

CALVING SPEED AT HEAD OF EMBAYMENT
COMPARED TO
DISCHARGE OF KNIK RIVER NEAR PALMER, ALASKA

Figure 8. Relation between rate of calving in Columbia Glacier embayment and the discharge of a nearby glacial river; from Sikonia and Post (1979).

Gilbert would no doubt have been pleased had he been able to see the results of a recent "monopulse radar" technique used in measuring the depth of Columbia Glacier (Fig. 6). Not only does it reveal that the ice extends as much as 470 m (1,550 ft) below sea level, but also that the glacier is in contact with its bed, not floating, throughout its length, so that active erosion at depth is possible. Indeed, the presence at the glacier's terminus of an underwater terminal-moraine shoal nearly 200 m high confirms that such active erosion is in progress.

Gilbert's observation that icebergs calve from tidewater glaciers more rapidly in summer than in winter has recently been confirmed; the rate of iceberg discharge from terminal embayments in Columbia Glacier has been found to be directly related to freshwater runoff (Fig. 8) (Sikonia and Post, 1979). The relationship suggests that during seasons of high runoff, high hydrostatic water pressure under the glacier may decouple the ice from its bed during brief periods, resulting in unusually rapid release of icebergs.

Although founded on physical principles that would be termed naive by present standards, his concept of the subglacial hydraulic system and its role in coupling the glacier to its bed is remarkably similar to the concepts that have emerged in the last decade. Glaciologists now consider that both water films and isolated conduits occur at the bed, and that this water controls the sliding. But the exact relation between water supply and sliding rate is perhaps the central problem in glacier dynamics.

Examining Gilbert's documents, the modern glaciologist may be surprised to discover that so many ideas, thought to be new, were in fact anticipated 80 yr ago by one of America's most remarkable and brilliant scientists. One can but wonder at what Gilbert would have accomplished if the sophisticated research tools now available had been available to him in 1899. Certainly it is fitting that one of the most esthetically pleasing and spectacular mountains in the area visited by the Harriman Expedition bears the name Mount Gilbert (see frontispiece).

REFERENCES CITED

Davis, William M., 1927, Biographical memoir of Grove Karl Gilbert, 1843–1918: Washington, D.C., National Academy of Sciences, v. XXI, Fifth Memoir, 303 p.

Field, W. O., 1932, The glaciers of the northern part of Prince William Sound, Alaska: Geographical Review, v. 22, no. 3, p. 361-388.
——1974, Glaciers of the Chugach Mountains, in Field, W. O., ed., Mountain glaciers of the Northern Hemisphere, Volume 2: Hanover, New Hampshire, Cold Regions Research and Engineering Laboratory, p. 299–492.
Gilbert, G. K., 1903, Glaciers and glaciation: (Alaska, Volume III) New York, Doubleday, Page & Company, 231 p. (Republished 1910, as Smithsonian Institution Harriman Alaska Series, Publication 1992.)
Grant, U.S., and Higgins, D. F., 1913, Coastal glaciers of Prince William Sound and Kenai Peninsula, Alaska: U.S. Geological Survey Bulletin 526, 75 p.
Merriam, C. Hart, 1919, Grove Karl Gilbert, the man: Sierra Club Bulletin, v. 10, no. 4, p. 391-399.
Nye, J. F., and Martin, P.C.S., 1968, Glacial erosion, in Ward, W., ed., General Assembly of Bern, Commission of Snow and Ice, Reports and Discussions: Gentbrugge, Belgium, International Association of Scientific Hydrology, Publication No. 79, p. 78-86.
Post, Austin, 1975, Preliminary hydrography and historic terminal changes of Columbia Glacier, Alaska: U.S. Geological Survey Hydrologic Investigations Atlas 559, 3 sheets.
Reid, Harry F., 1892, Report of an expedition to Muir Glacier, Alaska, with determinations of latitude and the magnetic elements at Camp Muir, Glacier Bay: Report of the Superintendent of the U.S.Coast and Geodetic Survey, p. 487-501.
——1896, Glacier Bay and its glaciers: U.S. Geological Survey Sixteenth Annual Report, pt. I, p. 415-461.
Sikonia, William G., and Post, Austin, 1979, Columbia Glacier, Alaska: Recent ice loss and its relationship to seasonal terminal embayments, thinning, and glacier flow: U.S. Geological Survey Open-File Report 79-1265, 3 sheets.
Tarr, R. S., and Martin, Lawrence, 1914, Alaskan glacier studies of the National Geographic Society in the Yakutat Bay, Prince William Sound and lower Copper River regions: Washington, D.C., National Geographic Society, 498 p.

MANUSCRIPT RECEIVED BY THE SOCIETY FEBRUARY 12, 1980
MANUSCRIPT ACCEPTED MAY 20, 1980

Geological Society of America
Special Paper 183
1980

Techniques and interpretation: The sediment studies of G. K. Gilbert

LUNA B. LEOPOLD
400 Vermont Avenue, Berkeley, California 94707

ABSTRACT

The laboratory experiments on sediment transport conducted by G. K. Gilbert differed importantly in technique from such studies of more recent date. Gilbert's flume was level and could not be altered in slope. Sediment was introduced at the upper end at a predetermined rate and by deposition built a bed gradient sufficient to transport the introduced load. The adjustment of slope in Gilbert's flume has contributed to the idea widely held by geologists that a river achieves equilibrium by adjusting its slope to provide just the velocity required for the transportation of the supplied load.

In fact, slope adjusts but little to a change in amount of introduced sediment load. The adjustment takes place principally among other hydraulic factors: width, depth, velocity, bed forms, channel pattern, and pool-riffle sequence. Gilbert sensed this complicated adjustment process, but its details are as yet only partially known in quantitative terms.

FLUME OPERATION

One of the most often referenced papers in the large literature on sediment transport is the famous U.S. Geological Survey Professional Paper 86 (Gilbert, 1914) in which Grove Karl Gilbert presented his data on the experiments in sediment transport. Interestingly, although this paper is probably quoted in references more often than nearly any other paper on the general subject, few people seem to have read it carefully. It is true that the analytical part of this long document is complicated, and the approach reflects the fact that at that time many things that are now taken for granted were not known.

The part that is probably not well understood concerns the implications of the methodology used by Gilbert in running the experiment. The most important aspect, and that which seems to be most overlooked, is the fact that his flume was built with a constant slope, a practice that differs from that of modern experimenters who use a more complicated flume in which the slope can be adjusted. The floor of Gilbert's flume was level, and the sideboards that formed the banks of the channel were high near the upstream end and decreased in steps toward the downstream end (see Pyne, this volume, Fig. 4B), Sand progressively built up in

the trough, thicker upstream than downstream, until the bed acquired a slope sufficient for the transport of introduced sediment. Discharge was specified, and Gilbert controlled it by an orifice opening. The load also was specified; sediment was put in by gravity through a feed system where the sand was made mobile by having a small amount of water coming in on top of the sediment storage tank. The dependent variable was the slope of the debris bed.

As sediment came from the hopper, deposition occurred in the upstream end of the flume, and a wedge of sediment gradually extended itself downstream until the slope of the debris bed became more or less uniform. In Gilbert's own words:

Near the head of the trough sand was dropped into the water at a uniform rate, the sand grains being of approximately uniform size. At the beginning of an experiment the sand accumulated in the trough, being shaped by the current into a deposit with a gentle forward slope. The deposit gradually extended to the outfall end of the trough, and eventually accumulation ceased, the rate at which sand escaped at the outfall having become equal to the rate at which it was fed above. The slope was thus automatically adjusted and became just sufficient to enable the particular discharge to transport the particular quantity of the particular kind of sand. The slope was then measured. Measurement was made also of the depth of the current; and the mean velocity was computed from the discharge, width, and depth. [1914, p. 17]

Contrast this method of operation with that now used in the recirculating type of flume. In modern work, great emphasis is placed on achieving a constant depth of water down the length of the flume. As equilibrium conditions are approached, the traveling carriage on the flume sides is run up and down the length of the flume to see that the water surface becomes parallel to the flume rails. Under most conditions it can be assumed that when the water surface is parallel to the rails, the depth of sediment on the bed is uniform and the bed slope, therefore, is also parallel to both rails and the bottom of the trough.

Gilbert did not worry much about the parallelism of water surface and debris bed. In general, he took the slope of the debris bed to be also the slope of the water surface. As Gilbert explained, at the end of the measuring or timing period during which time the sediment was being trapped, the discharge was then stopped, and then "The slope of the channel bed is measured, and the sand caught during the period recorded by the watch is weighed" (1914,

p. 22). In other words, the slope that Gilbert published is for the most part the slope of the debris bed measured after the experiment was completed. He also pointed out that when the debris surface was rough, "it was usually graded before measurement by scraping from crests into adjacent hollows" (1914, p. 25).

Looking at the tabulation of the original data, one can see that the water-surface slope was measured in about one experiment out of ten; therefore, in nine out of ten cases the final published slope was the slope of the debris bed.

The recirculating flume commonly used in modern experiments on sediment transport is a design chosen to eliminate one of the really great disadvantages of the type of equipment used by Gilbert. In the Gilbert flume, when sediment was fed in at the upstream end at a known and chosen rate, the sediment was accumulating downstream in the collection device and had to be removed from the settling tank and physically carried upstream for it to be used again. This required much handling of sediment material, not only necessitating considerable storage space but facilities for drying and moving of sediment.

The recirculating flume eliminates these difficulties by having the sediment returned to the upstream end through the return pipe, but it also means that the sediment becomes a dependent rather than an independent variable.

The implication of this apparently modest difference in technique is immense. One might say that, in a broad sense, the sediment engineer and modern hydraulician have paid too little attention to the implications of the techniques used by Gilbert. By the same token, the geologist and geomorphologist have probably paid too much attention to the results of the Gilbert experiments without giving adequate attention to the implications of the technique itself. The reason for saying this is that the geologist, without really knowing it, has been led to apply in his mind the major results of the Gilbert experiments despite lip service to the interdependence of the variables. To be more specific, the geologist has tended to follow closely the reasoning summarized at length in the paper by J. Hoover Markin, who stated and expanded on the idea that slope of the stream is adjusted to accommodate the discharge and load. Specifically, he said that his study of the subject "tends to confirm the standard geologic view that streams readjust themselves to new conditions primarily by adjustments in slope, and only in minor degree by modification of the channel section" (1948, p. 508).

It is logical to reach this conclusion when one looks at the Gilbert experiments in that the discharge and load were specified or chosen in advance, and the depositing sediment on the bed of the flume built up a slope sufficient to transport the imposed load under the conditions of discharge and width chosen as the fixed parameters of the particular experiment.

FIELD EXAMPLE OF THE GILBERT FLUME IN ACTION

Practically no exactly similar situation exists in nature except for the one described by Leopold and Bull (1979) at the Loop of the San Juan River in Utah, an abandoned meander of the river. In that geologic accident, the abandoned loop of the river, having a very flat slope, built a wedge of sediment within the confines of the canyon until, as in the Gilbert flume, the slope was sufficient to carry the debris derived from the drainage area under the conditions determined by drainage area and climate.

Neither in the paper dealing with the Berkeley experiments nor in the paper on hydraulic-mining debris did Gilbert imply that he would expect the natural river to act in the same manner as he found in the flume. That is, he did not infer from the experiments that when the debris load was increased the slope would increase, as Mackin would explain. This is the problem of whether an increased load will cause aggradation or whether it will cause the stream to alter its slope. In the paper on the laboratory experiments, Gilbert expressed it in the following way:

Whenever and wherever a stream's capacity is overtaxed by the supply of debris brought from points above a deposit is made, building up the bed. If the supply is less than the capacity, and if the bed is of debris, erosion results. Through these processes streams adjust their profiles to their supplies of debris. The process of adjustment is called gradation; a stream which builds up its bed is said to aggrade and one which reduces it is said to degrade. [1914, p. 219]

Gilbert explained the process of a stream adjusting to a change in load in slightly different terms in the paper on hydraulic mining. He said:

If a stream which is loaded to its full capacity reaches a point where the slope is less, it becomes overloaded with reference to the gentler slope and part of the load is dropped, making a deposit. If a fully loaded stream reaches a point where the slope is steeper, its enlarged capacity causes it to take more load, and taking of load erodes the bed. If the slope of a stream's bed is not adjusted to the stream's discharge and to the load it has to carry, then the stream continues to erode or deposit, or both until an adjustment has been effected and the slope is just adequate for the work. [1917, p. 26]

Any change of conditions which destroys the adjustment between slope, discharge, fineness, and load imposes on the stream the task of readjustment and thus initiates a system of changes which may extend to all parts of the stream profile. The mining debris disturbed the adjustment of streams by adding to their load. Reclamation of levees disturbs it by increasing the flood discharge in certain parts of the river channels.

The law of adjusted profiles applies to streams with mobile beds—alluvial streams. Streams with fixed beds are normally underloaded. . . . This process works toward an adjustment of slopes, but with exceeding slowness, and the factors involved are different from those of alluvial streams. [1917, p. 27]

DOES A CHANGE IN LOAD CAUSE SLOPE TO CHANGE?

The crux of the matter is the extent to which slope will alter if sediment load changes. The near parallelism of terraces with present streambeds implies that, in most situations, change in load-discharge relationships in a given basin results in aggradation or degradation with nearly no change in channel slope. In those cases where a change of slope, as well as aggradation or degradation, is observed, it will usually be also associated with a change in the size or size distribution of the sediment load.

The deposition of a valley fill or aggradation through a long reach of channel results from the introduction into the channel system of more sediment than can be transported through the system. The gradient of the deposited fill is the result of the interaction of the adjustable or dependent variables in the hydraulic process: width, depth, velocity, and the elements contributing to hydraulic resistance (roughness). These include sediment size and size distribution, bed forms, channel slope and pattern, bank undulations, curvature, and pool-riffle formation.

In discussing the laboratory experiments, Gilbert made a

separate analysis of the effect of various hydraulic parameters on capacity to carry load. The factors include discharge, slope, velocity, depth, form-ratio, and roughness, the roughness being treated less completely than any of the others. Gilbert certainly did not imply that any of these parameters could, in a natural stream, be treated separately, and his whole discussion indicated that he visualized an interdependence in which all were being mutually adjusted, just as Mackin stated. Mackin, however, made the further assertion, not to be found in Gilbert's work, that slope was the primary manner in which a stream adjusted to changed conditions, and that the effect of the adjusted slope was felt primarily in velocity, which Mackin stated was the principal determinant of the stream's capacity to carry sediment load.

RELATION OF GILBERT'S IDEAS TO MORE RECENT FINDINGS

Three important ideas have developed since the Gilbert experiments of which, of course, he had no knowledge. The first was the result of the Shields experiments, in which a relationship is expressed between sediment size and flow characteristics at the beginning or initiation of motion.

The second main idea is the hydraulic geometry: in any stream cross section, width, depth, and velocity vary with increasing discharge in a specific and conservative way. Similarly, as discharge changes downstream, the width, depth, and velocity change in a progressive manner.

The third main idea toward which Gilbert was reaching but did not at that time see is that the various dependent factors adjust mutually toward a most probable state—the concept of minimum variance.

However, when one looks at the present major computational scheme for estimating sediment load, the principal factor, now fairly well agreed upon by most workers, is stream power. This was well known to Gilbert, but its importance was apparently not completely apprehended by him. That he considered stream power important is indicated by the fact that it appears in the abstract of his paper. He said:

The energy of a stream is measured by the product of its discharge (mass per unit time), its slope, and the acceleration of gravity. In a stream without load the energy is expended in flow resistances, which are greater as velocity and viscosity are greater. Load, including that carried in suspension and that dragged along the bed, affects the energy in three ways. 1) It adds its mass to the mass of the water and increases the stock of energy pro rata. 2) Its transportation involves mechanical work, and that work is at the expense of the stream's energy. 3) Its presence restricts the mobility of the water, in effect increasing its viscosity, and thus consumes energy. For the finest elements of the load the third factor is more important than the second; for coarser elements the second is the more important. For each element the second and third together exceed the first, so that the net result is a tax on the stream's energy. Each element of load, by drawing on the supply of energy, reduces velocity and thus reduces capacity for all parts of the load. This principle affords a condition by which total capacity is limited. Subject to this condition a stream's load at any time is determined by the supply of debris and the fineness of the available kinds. [1914, p. 11]

Gilbert was ahead of his time in several important respects. His experimental technique, when viewed in terms of modern hydraulic laboratories, was simple. The present generation apparently has not benefited from this early example of simplicity, for the complications of current experimental techniques have added less

than one might suppose. No other experimenter has been willing to take the time to deal with such a large variety of sediment sizes as did Gilbert, and the fact that Gilbert's experiments have not been duplicated and extended now poses one of the serious problems in availability of laboratory data.

Furthermore, most modern experimenters have not yet seen the theoretical difference between the results obtained from recirculating flumes and those obtained from a sediment-feed flume such as Gilbert used. Until the work of Maddock (1969) and Langbein (1964), it was not visualized that the results would be slightly different if the sediment was fed as an independent variable or if the sediment load was a dependent variable as in a recirculating flume. In fact, most hydraulicians at the present time have not become convinced that this difference exists, but both theory and practice indicate that the results depend on which parameters are dependent and which are independent.

Gilbert was unwilling to estimate how, in a natural stream, these various hydraulic and sediment factors interact to produce the actual stream conditions. In this respect Mackin's (1948) well-known paper took a step beyond what Gilbert was willing to take, and in retrospect it was a misleading one. Mackin's emphasis on the change of slope in response to a change in load has misled geologists for three decades. Until the concept of minimum variance was put forward, the geologist accepted, apparently without question, that slope governs velocity and velocity governs sediment load; new work indicates that that is correct neither in theory nor in practice.

One of the problems Gilbert cited still remains as a major block to the advance of knowledge of natural streams, that is, the availability of sediment. The new measurements of bedload in natural streams made with the Helley-Smith sampler indicate that the results are highly dependent upon the sediment available to the stream for transport (Emmett, 1976, p. 12).

Thus the same problems enunciated by Gilbert still confront us: the amount and type of adjustment that occur among the individual parameters in a stream as a result of a change in imposed conditions. Gilbert grappled with the problem of the interrelationship between velocity, depth, slope, and discharge but was apparently not sensitive to the relationship between these and hydraulic resistance. This quandary is illustrated in the following discussion:

When depth is increased without change of slope (or width or grade of debris), its increase is effected by increase of discharge, with the result that capacity is increased, so that capacity is an increasing function of depth. When depth is increased without change of discharge, its increase is effected by reducing slope, with the result that capacity is reduced, so that capacity is a decreasing function of depth. When depth is increased without change of velocity, its increase requires increase of discharge accompanied by diminution of slope; and as these changes have opposite influences on capacity, it is not evident a priori whether capacity will be enlarged or reduced. The experimental data show that it is slightly reduced, so that capacity is a decreasing function of depth.

When depth is reduced without change of slope, and the reduction is continued progressively, a stage is eventually reached in which the velocity is no longer competent for traction. . . .

Reduction of depth without change of discharge involves increase of velocity, and it is evident that competence does not lie in that direction. But increase of depth involves reduction of velocity, and leads eventually to a competent velocity. The limiting depth corresponding to competence is therefore a great depth instead of a small one. . . .

When depth is reduced without change of mean velocity, the efficiency of the mean velocity is enhanced and competence is not approached. When depth is increased, the efficiency of the unchanged mean velocity is diminished and a [large] competent depth may, under some conditions, be realized. [1914, p. 166]

PROBLEMS OF THE PRESENT

These are the types of questions for which field data are needed to estimate quantitatively the reaction of a natural river system to changes imposed by man's activities. And the effect of man is the major river problem of the present day, just as it was in Gilbert's time.

Let us assume that as a result of man's activities, the incoming sediment load to a reach of river is altered as in the case of hydraulic mining. The work of Leopold and Bull (1979) suggested that slope is going to be the least adjustable of the factors. This stands in contrast to the conditions imposed by the construction and operation of the Gilbert flume. In the latter case the slope was the principal manner in which the adjustment took place, and the slope was altered until it was competent to carry the load that was imposed. The amount of adjustment made in velocity, depth, and roughness to such new conditions is as yet unknown, and new data are needed. After these adjustments have been made to the new conditions, the stream will aggrade if the load cannot be carried, or if the available load is insufficient to fulfill the needs of the adjusted conditions, the stream will degrade. If the size distribution of sediment remains constant, the aggradation and the degradation will probably take place with little or no change in river slope.

What slope a given reach of river will adopt apparently depends primarily on discharge and size of bed material but is also influenced by other factors "such as channel cross section and amount of load (Hack, 1957, p. 59)." But because there are not merely two but several factors interacting, the empirical relation between slope, bed material size, and discharge, such as that of Hack (1957, eq. 2), cannot forecast how a channel reach will react if the amount of introduced load is changed. Present knowledge admits merely that there are many interacting factors that mutually adjust toward a condition of dynamic equilibrium. This has been variously described, but perhaps best by Hack (1960).

The concept (of dynamic equilibrium) requires a state of balance between opposing forces such that they operate at equal rates and their effects cancel each other to produce a steady state, in which energy is continually entering and leaving the system (p. 86).

It is assumed that within a single erosional system all elements of the topography are mutually adjusted so that they are downwasting at the same rate (p. 85).

Hack considered this dynamic equilibrium as alternative to and different from the concept of the graded stream. He preferred to use the word "grade" or "graded" to refer to the smooth curve or smooth profile of any stream or reach of stream.

Leopold and Bull (1979) described a graded stream as follows:

A graded stream is one in which over a period of years slope, velocity, depth, width, roughness, pattern and channel morphology delicately and mutually adjust to provide the power and the efficiency necessary to transport the load supplied from the drainage basin without aggradation or degradation of the channels. The threshold of critical power is passed and the stream is not graded when the volume of load supplied is insufficient or is too large to be transported and the channel bed degrades or aggrades.

Hack considered this definition too restricted to have value because many streams have beautifully regular profiles but, in his opinion, would not fit the above definition. An example is the Middle River, Virginia, which he has shown to be littered with boulders plucked from the bed and therefore, he argues, must be degrading.

But neither the concept of dynamic equilibrium nor any definition of a graded stream indicates even qualitatively the degree to which slope adjusts in response to a change in sediment load.

The slowly accumulating field data suggest that the load supplied by the drainage basin affects a reach of channel both by its volume-per-unit time and by its size and size distribution. It appears that if size does not change but volume changes, the effect on channel slope will be minor, and aggradation or degradation will reflect the volume change. If size and its distribution are changed, the channel will tend to alter its slope by deposition and erosion. Debris size appears to have its principal influence through its effect on hydraulic roughness.

Yet these statements are not expressed in such quantitative terms that they can be used to forecast the results of changes now being quickly wrought by man's alterations of natural river systems. Thus, the slight improvement in concepts that has been made in the six decades since Gilbert's work is insufficient to solve present problems. A comprehensive program of field study comparable to the Gilbert approach to laboratory data is vitally needed.

REFERENCES CITED

Emmett, W. W., 1976, Bedload transport in two large, gravel-bed rivers, Idaho and Washington: Proceedings, Third Federal Inter-Agency Sedimentation Conference, Denver, Colorado, March 22–26, 1976.

Gilbert, G. K., 1914, The transportation of debris by running water: U.S. Geological Survey Professional Paper 86, 263 p.

——1917, Hydraulic-mining debris in the Sierra Nevada: U.S. Geological Survey Professional Paper 105, 153 p.

Hack, J. T., 1957, Studies of longitudinal stream profiles in Virginia and Maryland: U.S. Geological Survey Professional Paper 294-B, p. 45–97.

——1960, Interpretation of erosional topography in humid temperate regions: American Journal of Science, v. 258-A, p. 80–97.

Langbein, W. B., 1964, Geometry of river channels: Proceedings of the American Society of Civil Engineers, Journal of the Hydraulics Division, v. 90, HY2, p. 301–312.

Leopold, Luna B., and Bull, William B., 1979, Base level, aggradation, and grade; Proceedings of the American Philosophical Society, v. 123, no. 3, p. 168–202.

Mackin, J. H., 1948, Concept of the graded river: Geological Society of America Bulletin, v. 59, p. 463–512.

Maddock, T., Jr., 1969, The behavior of straight open channels with moveable beds: U.S. Geological Survey Professional Paper 622-A, 69 p.

Pyne, S. J., 1980, 'A great engine of research'—G. K. Gilbert and the U.S. Geological Survey, in Yochelson, E. L., ed., The scientific ideas of G. K. Gilbert: An assessment on the occasion of the centennial of the U.S. Geological Survey: Geological Society of America Special Paper 183 (this volume).

MANUSCRIPT RECEIVED BY THE SOCIETY OCTOBER 12, 1979
MANUSCRIPT ACCEPTED MAY 20, 1980

Geological Society of America
Special Paper 183
1980

G. K. Gilbert's geomorphology

R. J. CHORLEY
Department of Geography, Downing Place, Cambridge CB2 3EN, England

R. P. BECKINSALE
University College, Oxford, England

ABSTRACT

The scientific origins of Gilbert's geomorphic model are explored, and the distinctive character of that model is compared with that of the rival model of W. M. Davis, with special importance being given to the role of negative feedback by Gilbert. The early development of Gilbert's ideas (1871, 1875a, 1876) are treated, culminating in a detailed analysis of his classic work on the Henry Mountains (1877). His later work on coastal geomorphology, notably with reference to Lake Bonneville (1881, 1890) and on glacial geomorphology, showed a continuing preoccupation with the relations between form and process. An analysis of geomorphic work since World War II highlights the overriding importance of Gilbert's recent influence, particularly over the application of systems philosophy to the environmental sciences.

GILBERT AND DAVIS

In its third quarter the nineteenth century witnessed the advent of its two most important scientific paradigms, those of evolution and thermodynamics. It is not surprising that the developing scientific awareness of the two most important geomorphologists of that century should have been touched by these paradigms. It is well to remember that the year 1866 marked both the 16-yr-old Davis's independent observation of a new variable star (T Coronae; Chorley and others, 1973, p. 49) and the 23-yr-old Gilbert's scientific initiation with the Cohoes mastodon. The two young scientists were to profit very differently from these paradigms, however (see especially Pyne, 1975).

Davis, as his early predilection for entomology suggests, was most attracted to ideas stemming from theories of biological evolution. In his first exposition of the cycle of erosion (1884), he termed it a "cycle of life" and later wrote that "land forms, like organic forms, shall be studied in view of their evolution" (1905, 1909, p. 279; Chorley, 1965, p. 29-30). Ignoring, however, Darwin's major contribution concerning the *mechanisms* by which organic evolution may occur, Davis set about the study of landforms in terms of an evolutionary model that put the emphasis squarely on inevitable, continuous, and broadly irreversible processes of change viewed as producing an orderly sequence of landform transformations, wherein earlier forms could be con-

sidered as logical stages in a progression leading to later forms. Within this model, time became not a temporal framework within which events could occur but *a process itself* leading to an inevitable broad progression of change (Chorley, 1965, p. 30). Thus the passage of time, according to Davis, imprinted on landforms progressive changes in their geometry from which the former could be inferred as surely as from the changing geometric relations of the hands of a clock. It was this process view of time that led Davis to replace it with the term "stage" which, like other processes (the detailed nature of which he largely ignored), could proceed rapidly or slowly—unlike time! Although, in contrast with Gilbert, there is no direct evidence of Davis having studied thermodynamics, his time at the Harvard Lawrence Scientific School (1866-1870), which led to a Master's degree in engineering and to his permanent employment at that university after 1876, must have exposed him at least to its popular interpretations. Rudolf Clausius published his pioneer work on thermodynamics in the year of Davis's birth (1850) and formulated the two laws in 1865, including an explicit statement of the concept of entropy. J. Willard Gibbs returned from Europe to Yale in 1869 and began his influential publications on thermodynamics four years later, at a time when there was a public scientific dispute between Clausius and the followers of William Thomson (Lord Kelvin). The Second Law of Thermodynamics may have provided Davis with a less explicit, but nevertheless important, adjunct to his unidirectional view of evolution. The Second Law stated that in a closed system (that is, one incapable of receiving or giving off energy) entropy must increase irreversibly. Entropy is a measure of the extent to which energy within the system has ceased to be free, in the sense of being able to perform work on the system. A low-entropy closed system is one possessing within itself differentials in the distribution of energy such that the flow of energy from locations of high to low energy is capable of performing work. As time passes and the energy within the system becomes more equally distributed, the entropy increases until, at the state of maximum entropy, all parts of the closed system have the same energy levels, no flows of energy take place, and no work is being performed. Of course, as Davis was well aware, landform assemblages do not constitute closed systems and are constantly subjected to the import and export of energy. However, in the way that the model of the cycle of erosion was worked out by Davis between 1884 and 1899, the continuous energy inputs were either passed over or considered to parallel the increasing entropy levels within the system itself, leading to the maximum entropy state of the peneplain (Chorley, 1962). This view completely subordinated the general geomorphic effects of other energy inputs to the assumed dominant discontinuous inputs of potential energy due to relative uplifts between which the development of the landform system could be approximated by an entropy maximization or free energy decay model (Fig. 1A). The introduction of negentropy due to changes of base level were assumed by the model builder to be infrequent and to be followed by such a spatially uneven generation and temporally slow diffusion of free energy that the system was believed to retain at any one time a palimpsest of the effects of a series of superimposed energy inputs so that most landforms could be expected to exhibit a polycyclic character. These notions formed the basis of the science of denudation chronology that occupied the majority of world geomorphologists for many decades prior to World War II. Of course, Davis was too wise a general not to secure his rear, and in his 1899 exposition of the cycle of erosion, he allowed for a progressive reduction of inputs of orographic precipitation as the cycle progressed, although it is of note that this appears as yet another element in the general decrease of free energy as the cycle progresses, nor are the implications of such a precipitation decrease identified or worked

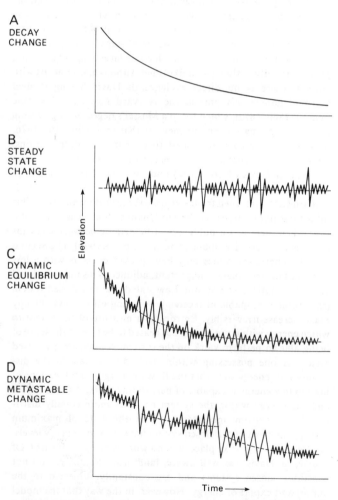

A
DECAY
CHANGE

B
STEADY
STATE
CHANGE

C
DYNAMIC
EQUILIBRIUM
CHANGE

D
DYNAMIC
METASTABLE
CHANGE

Elevation ⟶

Time ⟶

Figure 1. Four types of form change through time (after Schumm and Lichty, 1965; Chorley and Kennedy, 1971; Schumm, 1977). The changes shown are *schematic* only.

out in detail. Similarly, although some of the possible effects of erosion during long-continued slow uplift are suggested (Davis, 1905, p. 154), such a situation is considered to be rare and uplift to be generally relatively rapid. Amid the torrent of Davis's writings, eddies of every sort are incorporated, but they should not be confused with the coherent, remorsely unidirectional flow that constitutes the cycle of the erosion model. Davis placed such faith in the capacity and versatility of this model that he believed it to be a proper repository for assumed lesser models, two of which were the variable uplift model of the Pencks (Chorley and others, 1973, p. 520–528) and Gilbert's concept of grade (Chorley, 1962). A consideration of the latter permeates the present paper, and it is sufficient to note here that Davis employed grade in a rather different way than did Gilbert. Admittedly, Davis (1902, 1909, p. 399; see also Chisholm, 1967) allowed for the possibility of channel degradation after the achievement of the graded condition and for oscillations of cut and fill to accompany climatic changes, but the consequences of these possibilities were never worked out by him; such literary asides were entirely subordinated in his cyclic model to a general and progressive degradational decline in elevations and angles associated with the decrease in the amount and calibre of debris during the protracted time span of the cycle of erosion. In practice, Davis employed the concept of grade to identify a state which, being neither universal in space nor time, in its diffusion up river courses and valley-side slopes signaled the advent of his successive stages in landform history. For Gilbert, the assumption of a tendency for grade, ubiquitous in space and time, constituted the core of his theory of dynamic geomorphology.

Unlike Davis, Gilbert was relatively untouched by the theories of evolutionary biology, and it is significant that his first scientific work on the Cohoes mastodon led him on not to paleontology and taxonomy but to a study of potholing mechanisms and estimates of rates of waterfall recession (Gilbert, 1871). Davis regarded his own taxonomy of landforms, together with the accompanying terminology, as major personal achievement, whereas for Gilbert "rational classification" meant determining chains of necessary causal sequence (Gilbert, 1886, p. 285–286). On the other hand, it is known that Gilbert studied the thermodynamic work of Macquorn Rankine (Baker and Pyne, 1978, p. 104), and it can be argued that the Gilbertian geomorphic model was based on the energy conservation principle that formed the First Law of Thermodynamics and, in particular, upon the negative-feedback, homeostatic aspect of this which was to be formalized as a principle by Le Châtelier in the same year (1884) as Davis made his first tentative statement of the cycle of erosion (Chorley, 1962). This principle of self-regulation applied equally to both open and closed systems dominated Gilbert's view of the world. In this context it is perhaps not irrelevant to recall that Gilbert's grandfather invented a rotary steam engine in 1824 and his grandmother was the daughter of a noted clock maker (Mendenhall, 1920, p. 32). Whereas Davis had a linear view of causation (Fig. 2A), Gilbert's nonlinear vision was that "antecedent and consequent relations are. . .not merely linear, but constitute a plexus: and this plexus pervades nature" (Gilbert, 1886, p. 286). Gilbert's plexus (his sole biological analogy) constituted nested sets of negative-feedback loops intervening between the application of external processes and the production of the resulting adjusted equilibrium form (Fig. 2B) (see Bennett and Chorley, 1978, p. 111–115). For Davis, form evolved out of form; for Gilbert, the balance between externally applied processes and internally operative self-regulation found an outward and visible manifestation in adjusted topographic form. Gilbert viewed change in terms of homeostatic oscillations about equilibrium states, and it was natural that he should focus his attention on those aspects of landform development exhibiting what we should now term steady-state attributes (Fig. 1B). Baker and Pyne (1978, p. 104) have been at pains to point out the thermodynamic roots of Gilbert's thinking and that the work of J. Willard Gibbs on the equilibrium of chemical thermodynamic systems reached its culmination contemporaneously with Gilbert's research on the Henry Mountains. Just as Gibbs's chemical thermodynamic "phase" would readjust to a new one if any of the thermodynamic variables were changed, so Gilbert's fluvial equilibrium "grade" would readjust if any of the fluvial dynamic variables were changed. Thus Gilbert's open landscape systems were internally self-regulated toward equilibrium states in response to external process changes of mass and energy flows (Baker and Pyne, 1978, p. 114). As the same authors pointed out (1978, p. 113), Gilbert's steady-state model led to much confusion and misunderstanding on the part of Davis (Davis, 1927, p. 107–108; Chorley and others, 1973, p. 732–733) because Gilbert, who avoided any directional concept of geologic time, viewed time as a matrix for causes and a plexus of processes and employed time to express *rates* of change and *ratios* of force and resistance. In this approach Gilbert showed a strong engineering bias; and it is clear why he should have been attracted to the thermodynamic expositions of Rankine, who operated on the borderline between physical science and engineering (Pyne, 1975).

Gilbert's work on dynamic geomorphology had an immediate influence at home and in Europe. In the Old World his ideas on the equilibrium of river channels, slopes, and divides were welcomed as a strengthening of existing hypotheses. For example, in Germany, Philippson (1886) drew both on the work of French and Swiss civil engineers as well as on that of Gilbert when postulating that, under given conditions, a stable longitudinal profile of equilibrium (the *terminant*) would develop by working headward up the stream. The profile of this curve was held to be flat in humid

A. DAVIS

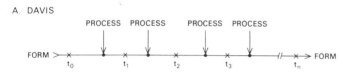

B. GILBERT

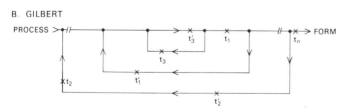

Figure 2. Diagrammatic representations of the contrasting Davis and Gilbert views of landform changes through time. For Davis (A) form changed in a linear manner with time ($t_0, t_1, t_2, t_3, ...t_n$). For Gilbert (B) the operation of nested negative-feedback loops of form on an input stream of process yielded various patterns of temporal change (for example, $t_1, t_2, t_3,$ $...t_n$ or $t'_1, t'_2, t'_3, ...t'_n$. Note the absence of t_0). Nested hierarchies of landforms at different spatial scales are controlled by a series of negative-feedback processes operating at various rates to produce a palimpsestic landscape, in which landforms exhibit differential temporal lagging in their response to the changing intensity of processes through time.

areas and steep in arid regions. In France, the important and influential work, *Les Formes du Terrain*, by the hydraulic engineer G. de la Noë and the geologist Emmanuel de Margerie (1888) reflected strongly the ideas on slope and fluvial processes enunciated by Gilbert, although they laid more stress on time and the control of base level (Chorley and others, 1964, p. 627–634). They asserted that present streams have

attained a state of equilibrium...; this equilibrium state (as clearly stated by Dausse in 1872) has special characteristics and deserves its name in the sense that in that state the bottom of the bed varies relatively little. [de la Noë and de Margerie, 1888, p. 56]

In their opinions on the erosion of the whole landscape, the authors express some Davisian views but acknowledge their debt to Gilbert, among others:

The ideas which have recently been expounded on the rôle of base-level, the upstream migration of incisions (nick-points), and the successive stages in the deepening of the bed, are... indicated, at least in substance, in several works. However, it seems that no one up to the present has had the idea of verifying them experimentally, as we have done. Among the authors who appear to have had the best understanding of the mechanics of fluvial erosion, we will mention Surell, Dana, Dausse, Gilbert, Heim, Dutton and Philippson. [de la Noë and de Margerie, 1888, p. 76]

Gilbert could hardly have wished for higher praise than to be included in such good company. However, his philosophy of dynamic geomorphology was soon to suffer a temporary, if prolonged, eclipse. Compared with the oversimplified model of Davis's cycle of erosion, which was explicitly constructed as a teaching device, Gilbert's geomorphic model was to a large extent implicit in his more extensive scientific monographs that were ostensibly concerned with more specialist matters. Indeed, the concept of equilibrium between force and resistance resulting in a clear formal response permeated all Gilbert's writings (Chorley and others, 1964, chap. 28; Baker and Pyne, 1978). Davis was a supreme teacher and advocate, whereas Gilbert was a modest and retiring field scientist whose brief teaching career lasted but a few months following his graduation from the University of Rochester in 1862. It is highly significant that, until the doctoral thesis of Pyne (1976), the only major critical biography of Gilbert was written by Davis (1927). The relative complexity of Gilbert's model, the relative obscurity of its exposition, his disdain of simple classificatory frameworks, his avoidance of any directional concept of geologic time (Baker and Pyne, 1978, p. 113), the illusion which Davis created of having successfully incorporated Gilbert's key ideas into the cycle of erosion, and, perhaps above all, a lack of student disciples meant that Gilbert's model was destined to be almost totally eclipsed until after the death of Davis. Only then did new generations discover in Gilbert's work a vitality which had progressively drained from the well-worn cycle of Davis. As Gilbert himself wrote: "The great investigator is primarily and pre-eminently a man who is rich in hypotheses. . . . The man who can produce but one, cherishes and champions that one as his own, and is blind to its faults" (1886, p. 287).

GILBERT'S FIRST GEOMORPHIC ESSAYS

In 1866 fragments of the skeleton of a mastodon were excavated in an abandoned pothole some distance below the Cohoes Falls on the Mohawk River. Gilbert, then a specimen curator, described the find in a local newspaper and met James Hall, the New York State Paleontologist, at the site. Subsequently, when Hall accidently slipped into the pothole and hurt himself, Gilbert supervised the remaining excavation. He immediately became interested in the potholes in this area, which he counted (350) and measured and described their morphometry (like a "chemist's test tube") and abrasional origin. The term "pothole" had been in use at least since 1839, and Archibald Geikie (1865, p. 28) had described its origin in an important work most possibly known to Gilbert. The latter then attempted to assess the rate at which the Cohoes Falls had retreated upstream, a problem long popular in America because of the notoriety of Niagara and other large waterfalls. Gilbert's method of dating was based on measurements of growth rings of cedar trees that had grown up on the gradually extending flat downstream of the retreating falls. From this dendochronology and the distances separating trees of different ages, he estimated the rate of recession to be about 12 in. (30 cm) per century (Gilbert, 1871). The minimum time for the retreat of the Cohoes Falls from the mastodon pothole to its present position was assessed at about 35,000 yr. These findings, the precise significance of which has been subsequently questioned, were less important than Gilbert's method—identify, measure, study frequency and morphometry, then with the aid of ratio analogues deduce causes and rates of operation (the dynamic *modus operandi*).

Gilbert (1889, 1895) remained interested in the retreat of waterfalls, as James Geikie (1894, p. 813) noted:

Mr. Gilbert, for example, concludes that the post-glacial gorge of Niagara, at the present rate of erosion, must have been cut out by the river within 7,000 years, while Mr. Winchell obtains for the post-glacial erosion of the Falls of St. Anthony (Minnesota) a period of 8,000 years. The observations and researches of Professor Spencer, however, show that the question of the age of the gorge of Niagara is a very complex one, and that Mr. Gilbert's estimate is probably much under the truth.

Passing over the controversy referred to in this comment, this quotation is important in drawing attention to the strong uniformitarian tendencies that were especially evident in Gilbert's earlier work. During later years Gilbert became more aware of geologic rhythms and of the importance of minor cataclysms, and to recognize that in a changing physical environment present and past rates of operation are not necessarily coincident, and that in dynamic geomorphology exact prediction based on short-term ratios and measurements are likely to be precarious unless the forms have already achieved dynamic equilibrium.

In three long field seasons with Lt. G. N. Wheeler's survey team between April 1871 and December 1873, Gilbert investigated much of Nevada, Arizona, southwest Utah, and western New Mexico, contributing extensively to the *Report* on these regions (Gilbert, 1875a). In chapter 1 he explained the topography of the Basin and Range area by a special type of block faulting, rather than in terms of the traditional hypothesis of erosion acting on folds. Chapter 2, entitled "Valleys, Cañons, Erosion," contains the germs of many ideas that were to be presented more forcibly in his monograph on the Henry Mountains. For example, Gilbert observed that in the Basin and Range province subaerial agencies have filled the original fault valleys, whereas on the Colorado Plateau region, rivers and rainwash had widened the canyons, and some valley-side cliffs had thus been worn back tens of miles (Gilbert, 1875a, p. 63, 68). In rivers the rapid torrent portion is deepening its bed and lowering its inclination; in the river proper the bed maintains a constant mean level and erosion widens the valley; in the delta the bed is rising by deposition (Gilbert, 1875a, p. 74-75). The torrent course will incise and flatten until "finally it can no longer clear its bottom of introduced detritus, and, its downward progress being arrested, the widening of its channel will begin" (Gilbert, 1875, p. 75). However, Gilbert's published work so far has little to distinguish it from that of Powell and in its emphasis on amount of strata denuded is virtually the same. He has recognized the great unconformity beneath the Paleozoic rocks of the Colorado Plateau, which he attributed to marine overlap; has discussed the dynamics of Basin and Range landforms and the formation of alluvial fans; and has made vague references to fluvial lateral planation and to a concept of grade. Gilbert was obviously ignorant of the important work that had been carried out by fluvial engineers in Europe, notably by Alexandre Surell (1841), and, more surprisingly, of the splendid writings of James Dwight Dana (1863), the first great American fluvialist (Chorley and others, 1964, p. 283-287 and 359-370).

Gilbert's next essay on landform dynamics (1876) made some significant advance on its predecessor, pointing out that the Colorado Plateau province, with its arid conditions, lack of vegetation and canyon dissection, was ideal for the revelation of the operation of structural and sedimentary systems. This article contained a foretaste of the brilliant chapter 5 of his Henry Mountains monograph (1877) and, in particular, previewed the notion of grade which, as shown by the following quotation, was significantly confused with "gradient":

In general, we may say that a stream tends to equalize its work in all parts of its course. Its power inheres in its fall and each foot of fall has the same power. When its work is to corrade and the resistance is unequal, it concentrates its energy where the resistance is great, by crowding many feet of descent into a small space; and diffuses it, where the resistance is small, by using but a small fall in a long distance. When its work is to transport, the resistance is constant, the fall is evenly distributed by a uniform grade. When the work includes both transportation and corrasion, as is the usual case, its grades are somewhat unequal; and the inequality is greatest when the load is least. [Gilbert, 1876, p. 100]

Later in life Gilbert (1914) modified his ideas on the relative importance of slope or gradient in controlling the graded condition in streams but, as we have seen, the concept of grade which Davis built into his cyclic synthesis was based significantly on Gilbert's earlier concepts. The assumed dominance of the slope effect also explains why Gilbert never felt it necessary to use the concept of base level which, for him, merely represented a terminally graded surface. As Davis (1924, p. 183) put it: "It is, I believe, not so much to Powell as to Gilbert that we owe the generally accepted idea of base level as a level base; in general, the extension of the ocean or geoid under the continents."

GILBERT'S MASTERPIECE ON THE DYNAMICS OF FLUVIAL PROCESSES

Gilbert first met John Wesley Powell with Clarence Dutton in Washington, D.C., in 1872 and in December 1874 joined Powell's staff. From then until November 1876 his field work was mainly on the Aquarius and Kaiparowits Plateaus and in the region of the Henry Mountains. The latter inspired his *Report on the Geology of the Henry Mountains* (1877) in which he based his whole explanation of terrain development on the dynamic adjustment of form to process. This matter had already been touched on by Powell himself (1876, p. 183-191), previewed by Gilbert (1876), and had been discussed at length with reference to hydraulic geometry by European engineers such as Louis Gabriel Du Buat (1786), M. F. B. Dausse (1857), Nathaniel Beardmore (1851), and T. J. Taylor (1851) (see Chorley and others, 1964, p. 88-90, 418-446). However, the Henry Mountains report contained the first major geomorphologic treatise on the subject by an earth scientist and, more important, an extension of the idea of process-form adjustments from erodible channels to whole landscapes.

In the important chapter 5 of his report, Gilbert deals first with the erosion of the Earth's surface and, in spite of some novel observations, adds little to what Powell had already written. Gilbert's (1877, p. 100-101) analysis of the disintegration of surface rocks is more thorough but does not come up to contemporary German standards (for example, Bischoff, 1846-1855). However, his discussion of fluvial geomorphic processes strikes a more authoritative note. Fluvial processes are shown to serve a treble purpose of removing eroded and weathered debris, of eroding the river bed, and of comminuting the transported material. Weathering, corrasion, and transportation are affected by many processes, variations in the ratios of which are critical. For example:

When the volume of a stream increases, it becomes at the same time more rapid, and its transporting capacity gains by the increment to velocity as well as by the increment to volume. Hence the increase in power of transportation is more than proportional to the increase in volume. [Gilbert, 1877, p. 104]

Or, again, in dealing with total rates of erosion:

Hence in regions of small rainfall, surface degradation is usually limited by the slow rate of disintegration; while in regions of great rainfall it is limited by the rate of transportation. There is probably an intermediate condition with moderate rainfall, in which a rate of disintegration greater than that of an arid climate is balanced by a more rapid transportation than consists with a very moist climate, and in which the rate of degradation attains its maximum. [Gilbert, 1877, p. 105. See also Powell, 1876, p. 188-189, and the important work by Langbein and Schumm, 1958, which the above quotation anticipates.]

Gilbert then examines the flow of a river, the total energy of which is measured by "the quantity of water and the vertical distance through which it descends." Although Hopkins (1842) had already shown that the maximum size of particle that can be moved by flowing water is proportional to the sixth power of the velocity (Beckinsale, 1972), "it must not be inferred that the total load of detritus that a stream will transport bears any such relation to the rapidity of its current "(Gilbert, 1877, p. 106). There is a difference between a stream's *competence* and its *capacity*, the latter being enhanced by debris comminution. Thus the velocity of a fully loaded stream depends (*ceteris paribus*) on the comminution of its load, and its velocity will be less when carrying its maximum load of coarse debris than when carrying its maximum load of fine detritus. If friction, load, and discharge are in harmony then slope is the main factor governing the rate of transportation, and "declivity favors transportation in a degree that is greater than its simple ratio" (Gilbert, 1877, p. 108). The whole argument continues with a rationale depending on the comparison of ratios, as in the following extract:

A stream's friction of flow depends mainly on the character of the bed, on the area of the surface of contact, and on the velocity of the current. When the other elements are constant, the friction varies approximately with the area of contact. The area of contact depends on the length and form of the channel, and on the quantity of water. For streams of the same length and same form of cross-section, but differing in size of cross-section, the area of contact varies directly as the square root of the quantity of water. Thus, *ceteris paribus*, the friction of a stream on its bed is proportioned to the square root of the quantity of water. But as stated above, the total energy of a stream is proportioned directly to the quantity of water; and the total energy is equal to the energy spent in friction, plus the energy spent in transportation. Whence it follows that if a stream change its quantity of water without changing its velocity or other accidents, the total energy will change at the same rate as the quantity of water; the energy spent in friction will change at a less rate, and the energy remaining for transportation will change at a greater rate.

Hence increase in the quantity of water favors transportation in a degree that is greater than its simple ratio. [Gilbert, 1877, p. 109]

The tendency to adjust transportation rates by means of changes of gradient to produce an "equilibrium of action" was shown to be ubiquitous:

Let us suppose that a stream endowed with a constant volume of water is at some point continuously supplied with as great a load as it is capable of carrying. For so great a distance as its velocity remains the same, it will neither corrade (downward) nor deposit, but will leave the grade of its bed unchanged. But if in its progress it reaches a place where a less declivity of bed gives a diminished velocity, its capacity for transportation will become less than the load and part of the load will be deposited. Or if in its progress it reaches a place where a greater declivity of bed gives an increased velocity, the capacity for transportation will become greater than the load and there will be corrasion of the bed. In this way a stream which has a supply of *débris* equal to its capacity, tends to build up the gentler slopes of its bed and cut away the steeper. It tends to establish a single uniform grade. [Gilbert, 1877, p. 112]

Here we must interrupt Gilbert's rationale to point out that the idea of a limiting slope of fluvial transportation had long been recognized by European hydraulic engineers, including Surell's *régime* (1841), Cunit's *courbe de régularisation* (1855) and Dausse's *pente de'équilibre* (1857). However, although Gilbert's account of river equilibration may have been plain fare for enlightened civil engineers, it was to become caviar for those workers with geomorphic tastes. This was achieved by expanding the concept of channel equilibrium to all humid transportational slopes and so embracing all humid landforms in his concept of dynamic equilibrium. This enormous jump was a stroke of genius and preceded a similar *tour de force* by W. M. Davis who, some decade later, was to leap from stages in river channel evolution to cycles of landscape development. But, as we have seen, the objectives of the two workers were strikingly different, as was their method of exposition. Gilbert, lacking the flowing prose and imaginative artistic ability of Davis, pursued his thesis as a Euclidean essay in plain English.

This rationale led Gilbert to evolve a number of major geomorphic theorems or laws, which extended notions of stream channel equilibrium to the associated valley-side slopes and visualized all fluvial processes as working in balance toward the disciplined and regulated reduction of the whole land surface. The *law of uniform slopes* stated:

It is evident that if steep slopes are worn more rapidly than gentle, the tendency is to abolish all differences of slope and produce uniformity. The law of uniform slopes thus opposes diversity of topography, and if not complemented by other laws, would reduce all drainage basins to plains. But in reality it is never free to work out its full results, for it demands a uniformity of conditions which nowhere exists. . . . The reliefs of the landscape exemplify other laws, and the law of uniform slopes is merely the conservative element which limits their results. [Gilbert, 1877, p. 115]

The *law of structure* postulated that:

Erosion is most rapid where the resistance is least, and hence as the soft rocks are worn away the hard are left prominent. The differentiation continues until an equilibrium is reached through the law of declivities. When the ratio of erosive action as dependent on declivities becomes equal to the ratio of resistances as dependent on rock character, there is equality of action. [Gilbert, 1877, p. 115-116]

The *law of divides* declared:

We have seen that the declivity over which water flows bears an inverse relation to the quantity of water. If we follow a stream from its mouth upward and pass successively the mouths of its tributaries, we find its volume gradually less and less and its grade steeper and steeper, until finally at its head we reach the steepest grade of all. If we draw the profile of the river on paper, we produce a curve upward and with the greatest curvature at the upper end. The same law applies to every tributary and even to the slopes over which the freshly fallen rain flows in a sheet before it is gathered into rills. The nearer the water-shed or divide the steeper the slope; the farther away the less the slope.

It is in accordance with this law that mountains are steepest at their crests. The profile of a mountain if taken along drainage lines is concave outward. . .; and this is purely a matter of sculpture, the uplifts from which mountains are carved rarely if ever assuming this form.

Under the *law of structure* and *law of divides* combined, the features of the earth are carved. Declivities are steep in proportion as their material is hard; and they are steep in proportion as they are near divides. [Gilbert, 1877, p. 116]

Gilbert envisaged these laws as providing opposing and balancing tendencies such that, for example, in soil- and vegetation-covered regions the law of divides may dominate over that of structure, whereas in arid areas the law of structure is supreme (Gilbert, 1877, p. 119). It was characteristic of Gilbert's thinking, however, that his most basic geomorphic principle was not separately identified as a "law" because it permeated the whole of his thesis:

The tendency to equality of action, or to the establishment of a dynamic equilibrium, has already been pointed out. . ., but one of its most important results has not been noticed. . . .in each basin all lines of drainage unite in a main line, and a disturbance upon any line is communicated through it to the main line and thence to every tributary. And as any member of the system may influence all the others, so each member is influenced by every other. There is interdependence throughout the system. [Gilbert, 1877, p. 123-124]

Having outlined his major thesis to the effect that landforms tend toward a state of dynamic equilibrium, even after interruptions to their formative processes, and that these fluvial processes form a spatial continuum, Gilbert went on to elaborate these ideas with respect to specific topographic features. He gave one of the first detailed descriptions of the formation of pediments by the planation of graded rivers:

Whenever the load reduces the downward corrasion to little or nothing,

lateral corrasion becomes relatively and actually of importance. The first result of the wearing of the walls of a stream's channel is the formation of a flood-plain. As an effect of momentum the current is always swiftest along the outside of a curve of the channel, and it is there that the wearing is performed; while at the inner side of the curve the current is so slow that part of the load is deposited. In this way the width of the channel remains the same while its position is shifted, and every part of the valley which it has crossed in its shiftings comes to be covered by a deposit which does not rise above the highest level of the water. The surface of this deposit is hence appropriately called the *flood-plain* of the stream. The deposit is of nearly uniform depth, descending no lower than the bottom of the water-channel, and it rests upon a tolerably even surface of the rock or other material which is corraded by the stream. The process of carving away the rock so as to produce an even surface, and at the same time covering it with an alluvial deposit, is the process of *planation*. [Gilbert, 1877, p. 126-127]

A few years later he wrote that the detritus worn from the mountain

is swept outward and downward by the flowing waters and deposited beyond the mouths of the mountain gorges. A large share of it remains at the foot of the mountain mass, being built into a smooth sloping pediment. . . . About the mouth of each gorge a symmetric heap of alluvium is produced—a conical mass of low slope, descending equally in all directions from the point of issue; and the base of each mountain exhibits a series of such alluvial cones. [Gilbert, 1881, p. 183]

However, in his analysis of alluvial cones he fell somewhat short of the important contemporary work of Dutton (1880, p. 219–221).

Gilbert considered river terraces to be the common result of river erosion acting upon an original flood plain. He stressed that they are not always predominantly built of debris but rather of rock in situ capped by debris and, as such, are largely the products of stream erosion, rather than of stream deposition. In this he differed widely from traditional ideas, for example, of Archibald Geikie (1865) in Scotland, and before him A. Smith (1832) in the Connecticut River valley. In the Henry Mountains, however, were terraces cut in solid rock which, incidentally, supported Gilbert's ideas on lateral planation by graded streams.

Another area of interest focused on the processes operative at stream heads (Gilbert, 1877, p. 121). In this respect Gilbert clearly anticipated both Horton's (1945) ideas of sheet flow runoff, merging into rills and then into discrete stream channels, as well as those of Hack and Goodlett (1960) on first-order valley-head processes. Elsewhere, Gilbert (1877, p. 141) discussed the monoclinal shifting of divides having asymmetrical slopes and the possible capture (abstraction) by more deeply eroding streams.

One of the striking features of Gilbert's report is the relative absence of reference to the ideas of his immediate superior, but in two instances, at least, the recently published notions of John Wesley Powell receive consideration. Firstly, Gilbert (1877, p. 126) allowed the possibility of antecedence by large rivers and, secondly, he greatly expanded, on genetic principles, Powell's (1875, p. 163, 165–166) classification of consequent, antecedent, and superimposed river valleys, thus:

1. Consequent upon original structure.

2. Antecedent to original structure after slow displacements.

3. Mixed antecedent and consequent after moderately rapid and rapid displacements.

4. Superimposed by sedimentation occurring after the submergence of an eroded landmass. After re-emergence the unconformable sediments impose a new drainage pattern upon the older series of strata.

5. Superimposed by alluviation, as, for example, by stream incision through an alluvial cone into the underlying rock.

6. Superimposition by planation:

The drainage of a district of planation is independent of the structure of the rock from which it is carved; and when in the progress of degradation the beds favorable to lateral corrasion are destroyed and the waterways become permanent, their system may be said to be *superimposed by planation.* [Gilbert, 1877, p. 144]

In terms of geomorphology, however, Powell's historical Colorado River and Uinta Mountains were poles apart from Gilbert's functional Henry Mountains, just as the emotional ruggedness of the former geologist contrasted so strikingly with the sensitivity of the latter.

COASTAL GEOMORPHOLOGY

From 1877 to 1879 Gilbert was mainly concerned with basic triangulation and irrigation problems in Utah but began his studies of Lake Bonneville in 1879. These continued, with breaks, for a further decade or so during which time he was largely concerned with administrative and supervisory work in Washington, D.C., where he served as Chief Geologist to the U.S. Geological Survey after 1888. His magnificent monograph on the Quaternary history of Lake Bonneville (1890) is generally regarded as his masterpiece, applying as it does Gilbert's notion of equilibrium to a wide range of associated problems from long-term crustal deformation to short-term beach processes. This monograph is given separate treatment elsewhere in this volume but it is necessary here to draw attention to Gilbert's important work on coastal geomorphology which appeared in his "Contributions to the History of Lake Bonneville" (1881), "Topographic Features of Lake Shores" (1885) and in chapter 2 of his Lake Bonneville monograph (1890).

Gilbert introduced his coastal studies with characteristic frankness, showing the freshness of approach that he always brought to problems in the field:

It happens, moreover, that the present treatment of the subject has its own peculiar point of view, and is in large part independent. During the progress of the field investigation I was unaware of the greater part of the literature mentioned above, having indeed met with but one important paper, that in which Andrews describes the formation of beaches at the head of Lake Michigan, and I was induced by the requirements of my work to develop the philosophy of the subject *ab initio.* The theories here presented had therefore received approximately their present form and arrangement before they were compared with those of earlier writers. They are thus original without being novel, and their independence gives them confirmatory value so far as they agree with the conclusions of others. [Gilbert, 1890, p. 25]

What resulted from Gilbert's observations was a unified general treatise on shore processes that to a large extent generated an interest in this branch of geomorphology culminating in Douglas Johnson's (1919) book on *Shore Processes and Shoreline Development.* Although Johnson took issue with Gilbert on the formation of offshore bars (barrier beaches), he made a significantly greater number of references to the work of Vaughan Cornish alone than to that of Gilbert. From his studies of the Lake Bonneville shore features, together with related studies, Gilbert was able to make important contributions regarding the formation of spits and barrier beaches, the equilibrium of beach processes, the

importance of infrequent events, the formation of ripples, and delta building. All these have, in a positive or negative sense, stimulated subsequent work in coastal geomorphology.

Spits and Barrier Beaches

Gilbert (1890, p. 40) did not clearly distinguish between spits and bars, on the one hand, and barrier beaches (offshore bars) on the other. Spits were believed to become barrier beaches when they extended for some distance parallel to the shore, bars when they linked together land or islands, and barrier islands when breached by storm waves (Gilbert, 1885, p. 87). They became recurved at the ends when wave erosion balanced rate of debris supply. These features were viewed as being produced near the wave zone along a line of agitation—"the road along which shore drift travels" (Gilbert, 1890, p. 40)—by material derived from the land, either from rivers or, mostly, from the cutting of marine "terraces" backed by cliffs (Gilbert, 1881, p. 172). Gilbert ascribed littoral transportation to a combination of wave action and currents generated by wind and tides. Although he believed that shore processes were impossible without the action of waves in setting material in motion, even obliquely attacking waves were viewed as relatively ineffectualy lateral transportation agents in the absence of dominant longshore currents (Gilbert, 1885, p. 85–86). Although Gilbert first allowed for water-level changes in assuming that submarine bars were submerged barrier beaches (Gilbert, 1885, p. 111), he later ascribed the features to subaqueous oscillations (1890, p. 44) and in general did not lean heavily on changes of water level in the production of constructional shore features. He believed that shoreline emergence resulted in erosion. These aspects of Gilbert's coastal geomorphology have proved the least durable in that subsequent work has put much greater stress on wave action as a debris transporting agent, on offshore sources of debris supply, on the importance of storm events of high magnitude, and on the effects of changes of sea level in the production of coastal deposition features. The last of these is still a matter of some contention, however, with followers of Johnson (1919) ascribing barrier beaches to negative movements of sea level and Hoyt (1967) to the submergence of ridges built up to the landward of the shoreline.

Equilibrium of Beach Processes

It was natural that one of Gilbert's central preoccupations should be with the establishment of equilibrium between the processes of beach erosion and deposition in an interdependent system with a short relaxation time and its manifestation in terms of the beach-face form.

The pulsating current of the undertow has an erosive as well as a transporting function. It carries to and fro the detritus of the shore, and, dragging it over the bottom, continues downward the erosion initiated by the breakers. This downwards erosion is the necessary concomitant of the shoreward progress of wave erosion; for if the land were merely planed away to the level of the wave troughs, the incoming waves would break where shoal water was first reached and become ineffective at the water margin. In fact, this spending of the force of the waves where the water is so shallow so as to induce them to break, increases at that point the erosive power by pulsation, and thus brings about an interdependence of parts. [Gilbert, 1885, p. 82–83]

In order that a particular portion of shore shall be the scene of littoral transportation, it is essential first that there be a supply of shore drift; second, that there be shore action by waves and currents; and in order that the local process be transportation simply, and involve neither erosion nor deposition, a certain equilibrium must exist between the quantity of the shore drift on the one hand and the power of the waves and currents on the other. On the whole this equilibrium is a delicate one, but within certain narrow limits it is stable. That is to say, there are certain slight variations of the individual conditions of equilibrium, which disturb the equilibrium only in a manner tending to its immediate readjustment. For example, if the shore drift receives locally a small increment from stream drift, this increment by adding to the shore contour, encroaches on the margin of the littoral current and produces a local acceleration, which acceleration leads to the removal of the obstruction. Similarly, if from some temporary cause there is a local defect of shore drift, the resulting indentation of the shore contour slackens the littoral current and causes deposition, whereby the equilibrium is restored. [Gilbert, 1890, p. 60–61]

One feels that Gilbert fell short of making a coherent statement of the concept of the beach profile of equilibrium, because he placed his emphasis on longshore debris movement principally by currents, rather than movement quasi-orthogonally to the shore effected by wave action. Recent work supports Gilbert.

High Magnitude Events

The importance of infrequent storm events in building up a beach was stressed by the belief that the direction of greatest shore transportation was that of maximum shore currents and largest storm waves, that wave-built terraces were due to storm wave action, and that each beach ridge was a record of some exceptional storm event (Gilbert, 1890, p. 42 and 55–57).

Formation of Ripples

Gilbert (1875b) described wave oscillation ripples as due to the frictional effect of running water, so confusing them with current ripples, but nine years later suggested a possible analogy between such ripple marks and the vibration of elastic bodies (1884b, p. 375–376). His work on the Lake Bonneville shorelines clarified his mind on this score and he later described giant symmetrical wave oscillation ripples in the Medina Sandstone of New York State (Gilbert, 1899, p. 135–140) and conducted experimental work leading to the production of a range of current ripples separated by thresholds (Gilbert, 1914).

Deltas

Except, possibly, for his suggestions regarding beach equilibrium, Gilbert's most lasting contribution to coastal geomorphology lay in his analysis of delta processes and forms. He outlined the classical delta theory whereby the discharge of water of equal density into a lake basin (that is, homopycnal flow) produced vertical mixing and sediment settling limited by the shallow depth of the basin of sedimentation (that is, plane jet flow). This generated a structure of dominantly foreset bedding with very subsidiary topset and bottomset beds. These findings were recently experimentally supported and theoretically expounded by Jopling (1963) and placed in the wider theoretical context of plane and axial flow of homopycnal, hypopycnal, and hyperpycnal type (Bates, 1953).

GLACIAL GEOMORPHOLOGY

Gilbert's contributions to glacial geology are dealt with elsewhere in this volume but it is appropriate here to say a word regarding his important, if limited, work on glacial geomorphology. This work stems largely from his visit to Alaska in 1899 as a member of the Harriman Expedition and from his researches in the Sierra Nevada a few years later. In many ways Gilbert's longest contribution (*Glaciers*, 1903, 231 p.) was the least stimulating. A great deal of it involved a general description of regional glacial features and the most interesting references were to the processes of glacial valley deepening and backcutting, together with the associated hanging valleys (1903, p. 114–122), cirques, and fjords. In connection with the latter, Gilbert (p. 210–218) postulated correctly that glaciers entering the sea will continue to exert the same pressure on their beds as will subaerial glaciers of the same thickness until the process of flotation actually begins. He also drew constant analogies between fluvial and glacial channel processes in their tendency to produce simple forms in longitudinal profiles, especially of their flow *surfaces* (p. 115) and in the importance of the conjunction of velocity and viscosity in erosion: "Most of the inequalities of velocity are determined by gravity in conjunction with the friction of the ice on the channel and the resistance of ice to internal shear; and the processes are essentially the same as with water" (Gilbert, 1903, p. 203). It is clear, however, that the prolonged relaxation times of glacial channel systems, compared with fluvial channels, severely restricted Gilbert's application of equilibrium notions to glacial terrain, and his Harriman report is of especial interest to the fluvial geomorphologist in that it contains Gilbert's major concession to Davisian geomorphology in the recognition of a high peneplain and several lower ones modified by ice action (Gilbert, 1903, p. 122–139). This conceptual difficulty meant that Gilbert's contribution to glacial geomorphology was destined to reside in detailed studies of local erosion processes. In 1904 he welcomed Willard Johnson's concurrently published hypothesis on cirque backcutting by freeze-thaw action in the bergschrund, named the schrund line (Gilbert, 1904a, p. 582) as a zone of back- and lateral-cutting in glacial troughs from which rock is quarried and exported, and used differential cirque backcutting to explain the steeper northern and eastern slopes of the Sierras associated with the asymmetry of glaciated mountains. Two years later Gilbert (1906a) described the production of hollows deepened by the potholing action of englacial water in a zone of ablation and also investigated the dynamics of crescentic-shaped gouges such as occur on the faces of some granitic bosses in the Sierra Nevada (1906b; Fig. 3). These gouges vary from a few centimetres to 2 m across and are arranged one behind the other in lines of two to six. They are outlined by fractures, some dipping at a gentle angle to the upper rock face and others, apparently the youngest, at a steep angle to it. The lune-shaped chip of rock between these fracture systems has been removed to form a crescentic gouge, with its convex steep-sided edge opposed to the former motion of the ice. T. C. Chamberlain (1888) thought this type of "chatter mark" resulted from a vibrating horizontal motion such as occurs when a cutting tool is forced across a hard surface. Gilbert accepted the idea that the rhythmic linear pattern was due to rhythmic movement of ice which progressed by the interstitial melting and regelation of its rigid crystalline grains. But he argued that the gouge was related to the "conoid of percussion" such as

Figure 3. Glacial striae and chattermarks on granite outcrop in the canyon of the Tuolumne River, Yosemite National Park, California. The ice motion was from right to left (U.S. Geological Survey).

would result from a sharp blow; that these gouges were formed by sizeable boulders in the basal ice which on meeting the obstruction of a projecting granite boss were pressed, with increased vertical pressure, upon its surface which was deformed elastically and then finally ruptured along a conoid fracture. However, because the local bedrock does not show signs of striation, Gilbert suggested that sand or some basal detritus cushioned the contact. In 1930 an adaptation of Gilbert's idea of crescentic gouges being the result of conoids of fracture gained wide credence (Ljungner, 1930) and, indeed, all his theories of glacial erosional processes still find wide support.

GILBERT'S LATER CONTRIBUTIONS TO FLUVIAL GEOMORPHOLOGY

Increasingly, Gilbert became more concerned with trying to unravel the detailed processes responsible for the production of equilibrium landforms. In 1884a he had tried to apply Ferrel's Law of Deflection to river channel meandering. That law, based on the Coriolis "force" resulting from the Earth's rotation (in an eastward direction), states that moving bodies are deflected to the right of their course in the Northern Hemisphere. Gilbert assumed that this effect might help to raise the water level on the right bank of a river and increase water pressure there so favoring a tendency to erode

that bank. Once the channel has become curved there is some competition between right and left banks, and Gilbert developed a formula of the ratio between the selective influences that will determine which bank will be favored.

$$\frac{\text{Right}}{\text{Left}} = \frac{V + pn \sin \text{lat}}{V - pn \sin \text{lat}}$$

where n = angular velocity of Earth's rotation, V = mean velocity of stream, and ρ = radius of stream curvature. When applied to the Mississippi, the formula indicated that the tendency to erode the right bank was nearly 9% greater than the tendency to erode the left bank.

Gilbert's interest in hydraulic geometry found abundant expression as the result of his pioneer flume experiments at Berkeley between 1907 and 1909—which were subsequently published in his important monograph on the transportation of debris by running water (1914)—as well as in his observations on the effects of hydraulic mining debris on certain California rivers carried on between 1905 and 1915 (1917). These works are being analyzed in detail elsewhere in this volume. Suffice it to say here that in his attempts to study the effects of four major factors (discharge, slope, fineness, and depth/width form ratio) on bed-load capacity, Gilbert introduced the concept of thresholds (that is, competent slope, discharge, and fineness) and showed that changes in capacity produced predictable changes in stream-bed geometry. One of the most innovative aspects of this work was his recognition of the problems of dealing with multivariate, nonlinear systems. Of his failure truly to isolate the effects of single factors controlling bed-load capacity, Gilbert wrote (1914, p. 109): "The development of complexity within complexity suggests that the actual nature of the relation is too involved for disentanglement by empiric methods," adding enigmatically "but that conclusion does not necessarily follow." Similarly, in relating capacity to mean velocity it was "necessary to postulate constancy in some accessory condition" (1914, p. 10–11). In this work Gilbert confronted the plexus more squarely than in any other of his endeavors and, in so doing, previewed the multivariate methods of Melton (1957, 1958a, 1958b) as well as the indeterminacy ideas of Langbein and Leopold (1964).

In his work on the Henry Mountains some 30 yr previously Gilbert had been puzzled by an apparent violation of the law of divides by convex badland crests (Fig. 4):

Evidently some factor has been overlooked in the analysis, a factor which in the main is less important than the flow of water, but which asserts its existence at those points were the flow is exceedingly small, and is there supreme. [Gilbert, 1877, p. 123]

Following the intervening work by Davis (1892), Gilbert now recognized the importance of creep in a location where it is dominant over surface runoff processes. In addition, Gilbert's short paper on the convexity of hilltops (1909) carried two important extensions of his thinking on dynamic geomorphology. Firstly, he attempted to reason from a pure vision of process to an unambiguous resulting topographic form (1909, p. 345–346). Despite his lack of complete success, which was paralleled in his subsequent work on channel processes (1914), this attempt reiterated his faith in an open system view of landform development. For him the closed system statement "movements of material tell the earth's surface how to slope and slope tells material how to move" was not the whole story. Secondly, his law of creep (1909, p. 345) is a statement that transcends mere mass movements. Like a chain reaction this law demolishes the distinction between mass movements, on the one hand, and weathering and all other fluvial processes, on the other. Just as an earlier paper Gilbert (1904b) had linked exfoliation weathering in the Sierra Nevada with long-preceding tectonic processes (Fig. 5), so his Henry Mountains monograph had implied that weathered material is the parent of most landforms with geology being relegated to grandparenthood!

We have seen that vegetation favors the disintegration of rocks and retards the transportation of the disintegrated material. Where vegetation is profuse there is always an excess of material awaiting transportation, and the limit to the rate of erosion comes to be merely the limit to the rate of transportation. And since the diversities of rock texture, such as hardness and softness, affect only the rate of disintegration (weathering and corrasion) and not the rate of transportation, these diversities do not affect the rate of erosion in regions of profuse vegetation, and do not produce corresponding diversities of form. [Gilbert, 1877, p. 119]

For Gilbert landscape processes were a unity, landforms a spatially nested continuum (Fig. 6). Davis looked out over the folded Appalachians and saw a palimpsest resulting from the historical superimposition of the effects of cycles and parts of cycles; Gilbert saw in Utah a dynamic process-response surface.

THE INFLUENCE OF GILBERT

For some 30 yr after his death Gilbert's geomorphic ideas, like those of Powell and W. Penck, were interpreted and found acceptance in the English-speaking world mainly through the works of W. M. Davis and his adherents. Even the important paper by Robert E. Horton (1945), which in many respects marked the beginning of modern quantitative geomorphology, made no mention of the work of Gilbert in the body of the text, despite the inclusion of two of his works in the references. The reasons for this temporary eclipse have already been touched upon and it is well to note that Gilbert was temperamentally much more of a researcher than a teacher or proselytizer. His one textbook (Gilbert and Brigham, 1902), although extremely competent on geomorphic processes and geological influences on terrain, was distinctly conventional and contained no mention of the concept of grade whatsoever.

However, the decline in influence of the ideas of Davis after World War II (Chorley, 1965), together with a rise in popularity of functional studies relating form to process (Chorley, 1967; 1978, p. 7–9), led to a revival of interest in the first-hand contributions of Gilbert, which revival is still dominant in dynamic geomorphology (Baker and Pyne, 1978). Mackin's (1948) return to the negative-feedback analogy of Le Châtelier's thermodynamics gave the first important 20th century elaboration of the concept of grade, and this was quickly followed by Wolman (1955) who showed the ability of alluvial rivers to grade to be more versatile and ubiquitous than Davis had interpreted, thereby sweeping away both the special temporal significance ascribed to grade by Davis and, paradoxically, the need to use the term at all! Gilbert's holistic view of landforms as an association of channels, slopes, and divides lay at the heart of the post–World War II revival of fluvial geomorphology, particularly in the works of Horton (1945) and Strahler (1950), and much of subsequent work has focused on these relationships (Schumm, 1956a; Chorley and Kennedy, 1971; Schumm, 1977) together with the preservation of these hierarchical forms through progressive time changes (Woldenberg, 1966, 1969; Bull, 1975). Naturally, the problem of landform changes through time has continued to dominate geomorphology. The problem of explaining existing landform variety in terms of steady-state principles has revived (Bretz, 1962; Holmes, 1964; Abrahams, 1968; Ollier, 1968; Conacher, 1969), and Hack (1960) even had the temerity to throw a stone through the Davisian stained-glass window by attempting a steady-state interpretation of the hallowed polycyclic central Appalachians, an action followed in the equivalent British Wealden shrine by Worssam (1973). Such steady-state essays have tended to be united with Davisian concepts of irreversible change to produce dynamic equilibrium models of landform change through time (Fig. 1C). Schumm and Lichty (1965) suggested that Gilbertian, Davisian, and other time scales could be separately identified and associated with different types of geomorphic investigation, but in the last decade or so efforts have increasingly turned to the building of complex models appropriate to a wide range of time scales and exhibiting both steady-state and progressive decay characteristics. Further than this, another concept of Gilbert is being productively reinvestigated. Gilbert believed in the importance of small cataclysms in shaping landforms, which would subsequently equilibrate with respect to more frequent processes (Baker and Pyne, 1978, p. 113), and attention is now turning to the study of dynamic metastable changes (Langbein and Leopold, 1964; Chorley and Kennedy,

Figure 4. Badland topography with rounded divides developed on decomposed granite, exposed by hydraulic mining at North San Juan, Nevada County, California (U.S. Geological Survey photograph taken by G. K. Gilbert).

Figure 5. G. K. Gilbert by a large erratic granite boulder on the slope of Moraine Dome above Little Yosemite Valley, California, in 1908. The pedestal on which the erratic stands is the postglacially weathered remnant of a bedrock exfoliation shell (U.S. Geological Survey photograph taken by E. C. Andrews).

Figure 6. Aerial view of the Gilbert badlands northwest of the Henry Mountains, Utah, showing the organized hierarchy of nested drainage basins (photograph courtesy of Luna Leopold).

porary dynamic geomorphology. Melton (1957, 1958a, 1958b), using a Gibb's type phase space model, treated slopes and drainage systems from the viewpoint of multiple causation so dear to Gilbert. Schumm (1956a, 1956b) employed badlands research to generate general geomorphic principles in a manner worthy of Gilbert. The recent flood of research into hydraulic geometry (for example, Leopold and Maddock, 1953; Leopold and Wolman, 1957; Langbein and Leopold, 1966, 1968), particularly that employing flumes (Langbein and Wolman, 1957; Schumm and Khan, 1972), derives directly from the impetus given by Gilbert. Schumm (1963), drawing directly upon Gilbert's interests in measurements of process and in recent vertical movements of the Earth's crust, produced the most notable application of published process rates to a recurrent problem in classical geomophology since the important paper by Geikie (1868), almost a century before. Thermodynamic analogies abound (Leopold and Langbein, 1962; Scheidegger, 1964, 1967; Scheidegger and Langbein, 1966; Yang, 1971).

Above all, however, and in a very real sense uniting the foregoing, Gilbert's legacy to contemporary dynamic geomorphology resides in his anticipation of the systems approach to the discipline. Strahler's (1950) utilization of systems theory, which derived from the science of thermodynamics via Von Bertalanffy's biology, ushered in a true revival of Gilbertian thinking in geomorphology. Identified in more detail by Chorley (1962) and elaborated in its implications by Howard (1965), King (1967), King (1970), and Rayner (1972), the systems approach has of more recent years begun to confound its critics (for example, Chisholm, 1967; Smalley and Vita-Finzi, 1969) and to give the kind of coherence to geomorphology toward which Gilbert was only able to aspire (Chorley and Kennedy, 1971). In the past few years, systems theory has been productively applied to weathering (Trudgill, 1977), mass movements (Brunsden, 1973), sediment transport (Allen, 1974), and arroyo cutting (Cooke and Reeves, 1976); this shows not only the rich future possibilities for such an approach but also highlights existing needs in experimental design and field-data generation. More than this, the systems philosophy that the work of Gilbert so clearly anticipated (Baker and Pyne, 1978, p. 105) is emerging as the viable universal basis for modeling in the earth, atmospheric, oceanic, and biological sciences. This allows their interfacing with the socioeconomic decision-making systems and promises for the first time a unified and rational approach to man's environmental problems (Bennett and Chorley, 1978).

1971, chap. 6) which are characterized by the existence of thresholds and episodic changes (Fig. 1D; Schumm, 1973, 1975; Dury, 1975). Just as the last quarter of the 19th century saw the qualitative statements of the decay and steady-state models in geomorphology, the last quarter of the twentieth century may well be dominated by their quantitative union in the dynamic metastable model. However, it is clear that Gilbert's paramount concern with landscape *form*, as the manifestation of process, must remain the central theme of modern dynamic geomorphology (Strahler, 1952), and that geomorphologists will do well to cultivate his interests in models relating to the origin, maintenance, and transformation of landforms at appropriate middle-order scales.

In other, related ways, Gilbert's influence pervades contem-

REFERENCES CITED

Abrahams, A. D., 1968, Distinguishing between the concepts of steady state and dynamic equilibrium in geomorphology: Earth Science Journal, v. 2, p. 160–166.

Allen, J. R. L., 1974, Reaction, relaxation and lag in natural sedimentary systems: General principles, examples and lessons: Earth-Science Reviews, v. 10, p. 263–342.

Baker, V. R., and Pyne, S., 1978, G. K. Gilbert and modern geomorphology: American Journal of Science, v. 278, p 97–123.

Bates, C. C., 1953, Rational theory of delta formation: American Association of Petroleum Geologists Bulletin, v. 37, p. 2119–2161.

Beardmore, N., 1851, Manual of hydrology: London, Waterlow and Sons, 384 p.

Beckinsale, R. P., 1972, William Hopkins: Dictionary of scientific biography: New York, Charles Scribner's Sons, v. 6, 502–504.

Bennett, R. J., and Chorley, R. J., 1978, Environmental systems: Philosophy, analysis and control: London, Methuen, 624 p.

Bischof, C. C. G., 1846–1855, Lehrbuch der chemischen und physikalischen Geologie: Bonn, 3 v. (1863–1871, second edition; 1854–1859, London, 3 v., translated as Elements of chemical and physical geology).

Bretz, J H., 1962, Dynamic equilibrium and the Ozark land forms: American Journal of Science, v. 260, p. 427–438.

Brunsden, D., 1973, The application of system theory to the study of mass movement: Geologica Applicata e Idiogeologia (Bari), v. 7, p. 185–207.

Bull, W. B., 1975, Allometric change of landforms: Geological Society of America Bulletin, v. 86, p. 1489–1498.

Chamberlin, T. C., 1888, The rock scorings of the great ice invasions: U.S. Geological Survey, Seventh Annual Report (1885–1886), p. 155–248.

Chisholm, M.D.I., 1967, General systems theory and geography: Transactions of the Institute of British Geographers, no. 42, p. 45–52.

Chorley, R. J., 1962, Geomorphology and general systems theory: U.S. Geological Survey Professional Paper 500-B, 10 p.

——1965, A re-evaluation of the geomorphic system of W. M. Davis, in Chorley, R. J., and Haggett, P., eds., Frontiers in geographical teaching: London, Methuen, p. 21–38.

——1967, Models in geomorphology, in Chorley, R. J., and Haggett, P., eds., Models in geography: London, Methuen, p. 59–96.

——1978, Bases for theory in geomorphology, in Embleton, C., Brunsden, D., and Jones, D.K.C., eds., Geomorphology, present problems and future prospects: Oxford University Press, p. 1–13.

Chorley, R. J., and Kennedy, B. A., 1971, Physical geography; A systems approach: London, PrenticeHall, 370 p.

Chorley, R. J., Beckinsale, R. P., and Dunn, A. J., 1973, The history of the study of landforms, Volume 2, The life and work of William Morris Davis: London, Methuen, 874 p.

Chorley, R. J., Dunn, A. J., and Beckinsale, R. P., 1964, The history of the study of landforms, Volume 1, Geomorphology before Davis: London, Methuen, 678 p.

Conacher, A. J., 1969, Open systems and dynamic equilibrium in geomorphology; A comment: Australian Geographical Studies, v. 7, p. 153–158.

Cooke, R. U., and Reeves, R. W., 1976, Arroyos and environmental change in the American South-West: Oxford University Press, 213 p.

Cunit, C., 1855, Etudes sur les cours d'eau à fond mobile: Grenoble.

Dana, J. D., 1863, Manual of geology: Philadelphia, Theodore Bliss, 798 p.

Dausse, M.F.B., 1857, Note sur un principe important et nouveau d'hydrologie: Comptes Rendus de l'Académie des Sciences, Paris, v. 44, p. 756–766.

Davis, W. M., 1884, Gorges and waterfalls: American Journal of Science, 3rd, ser., v. 28, p. 123–132.

——1892, The convex profile of bad-land divides: Science, v. 20, p. 245.

——1902, Base level, grade, and peneplain: Journal of Geology, v.10, p. 77–111.

——1905, Complications of the geographical cycle: Eighth International Geographical Congress, Washington, D.C., 1904, Report, p. 150–163.

——1909, Geographical essays: Boston, Ginn and Company, 777 p.

——1924, The progress of geography in the United States: Association of American Geographers Annals, v. 14, p. 159–215.

——1927, Biographical memoir of Grove Karl Gilbert, 1843–1918: National Academy of Sciences, Biographical Memoirs, v. 21, 5th Memoir, 303 p.

De la Noë, G. D., and De Margerie, E., 1888, Les formes du terrain: Paris, Imprimerie Nationale, 205 p.

Du Buat, L. G., 1786, Principes d'hydraulique: Paris, De l'Imprimerie de Monsieur, v. 1, 453 p.; v. 2, 402 p.

Dury, G. H., 1975, Neocatastrophism?: Brazilian Academy of Sciences Annals, v. 47, Suplemento, p. 135–151.

Dutton, C. E., 1880, Report on the geology of the high plateaus of

Utah: Washington, D.C., U.S. Geographical and Geological Survey of the Rocky Mountain Region, 307 p.

Geikie, A., 1865, The scenery of Scotland: London, Macmillan, 360 p.

——1868, On denudation now in progress: Geological Magazine, v. 5, p. 249–254.

Geikie, J., 1894, The great ice age (3rd edition): London, Stanford, 850 p.

Gilbert, G. K., 1871, Notes of investigations at Cohoes, with reference to the circumstances of the deposition of the skeleton of Mastodon... under direction of James Hall: New York State Museum of Natural History, Annual Report 21, p. 129–148.

——1875a, Report on the geology of portions of Nevada, Utah, California, and Arizona examined in the years 1871 and 1872: U.S. Geographical and Geological Surveys West of the 100th Meridian, v. 3, Geology, Part 1, p. 17–187.

——1875b, Ripple-marks (abs.): Philosophical Society of Washington Bulletin v. 2, p. 61–62.

——1876, The Colorado plateau province as a field for geological study: American Journal of Science, 3rd ser., v. 12, p. 1–27.

——1877, Report on the geology of the Henry Mountains: Washington, D.C., U.S. Geographical and Geological Survey of the Rocky Mountain Region, 160 p.

——1881, Contributions to the history of Lake Bonneville: U.S. Geological Survey, Second Annual Report, 1880–1881, p. 167–200.

——1884a, The sufficiency of terrestrial rotation for the deflection of streams: American Journal of Science, 3rd ser., v. 27, p. 427–432.

——1884b, Ripple-marks: Science, v. 3, p. 375–376.

——1885, The topographic features of lake shores: U.S. Geological Survey, Annual Report 5, p. 69–123.

——1886, The inculcation of scientific method by example, with an illustration drawn from the Quaternary geology of Utah: American Journal of Science, 3rd ser., v. 31, p. 284–299.

——1889, The history of the Niagara River: Commissioners of State Reservation at Niagara, Sixth Annual Report, p. 60–84 (Reprinted in Smithsonian Institution Forty-fifth Annual Report, p. 231–257).

——1890, Lake Bonneville: U.S. Geological Survey Monograph 1, 438 p.

——1895, Niagara Falls and their history: National Geographic Monograph, v. 1, p. 203–236.

——1899, Ripple-marks and cross-bedding: Geological Society of America Bulletin v. 10, p. 135–140.

——1903, Glaciers, (Harriman Expedition series, Volume 3,): New York Doubleday, Page and Company, 231 p. (reprinted in 1910 by the Smithsonian Institution, Washington, D.C., Publication 1992).

——1904a, Systematic asymmetry of crest lines in the high Sierra of California: Journal of Geology, v. 12, p. 579–588.

——1904b, Domes and dome structures of the high Sierra: Geological Society of America Bulletin, v. 15, p. 29–36.

——1906a, Moulin work under glaciers: Geological Society of America Bulletin v 17, p. 317–320.

——1906b, Crescentic gouges on glaciated surfaces: Geological Society of America Bulletin, v. 17, p. 303–316.

——1909, The convexity of hilltops: Journal of Geology, v. 17, p. 344–350.

——1914, The transportation of debris by running water: U.S. Geological Survey Professional Paper 86, 263 p.

——1917, Hydraulic-mining debris in the Sierra Nevada: U.S. Geological Survey Professional Paper 105, 154 p.

Gilbert, G. K., and Brigham, A. K., 1902, An introduction to physical geography: New York, D. Appleton and Company, 380 p. (2nd edition, 1904; 3rd edition, 1906).

Hack, J. T., 1960, Interpretation of erosional topography in humid temperate regions: American Journal of Science, v. 258A, p. 80–97.

Hack, J. T., and Goodlett, J. C., 1960, Geomorphology and forest ecology of a mountain region in the central Appalachians: U.S. Geological Survey Professional Paper 347, 66 p.

Holmes, C. D., 1964, Equilibrium and humid climate physiographic processes: American Journal of Science, v. 262, p. 436–445.

Hopkins, W., 1842, On the elevation and denudation of the district of the lakes of Cumberland and Westmoreland: Geological Society of London Proceedings, v. 3, p. 764–765.

Horton, R. E., 1945, Erosional development of streams and their drainage basins; Hydrophysical approach to quantitative morphology: Geological Society of America Bulletin, v. 56, p. 275–370.

Howard, A. D., 1965, Geomorphological systems—Equilibrium and dynamics: American Journal of Science, v. 263, p. 302–312.

Hoyt, J. H., 1967, Barrier island formation: Geological Society of America Bulletin, v. 78, p. 1125–1136.

Johnson, D. W., 1919, Shore processes and shoreline development: New York, John Wiley & Sons, Inc., 584 p.

Jopling, A. V., 1963, Hydraulic studies on the origins of bedding: Sedimentology, v. 2, p. 115–121.

King, C.A.M., 1970, Feedback relationships in geomorphology: Geografiska Annaler, ser. A, v. 52, p. 147–159.

King, R. H., 1967, The concept of general systems theory as applied to geomorphology: Albertan Geographer, v. 3, p. 29–34.

Langbein, W. B., and Leopold, L. B., 1964, Quasi-equilibrium states in channel morphology: American Journal of Science, v. 262, p. 782–794.

——1966, River meanders—Theory of minimum varance: U.S. Geological Survey Professional Paper 422-H, 15 p.

——1968, River channel bars and dunes—Theory of kinetic waves: U.S. Geological Survey Professional Paper 422-L, 20 p.

Langbein, W. B., and Schumm, S. A., 1958, Yield of sediment in relation to mean annual precipitation: American Geophysical Union Transactions, v. 39, p. 1076–1084.

Leopold, L. B., and Langbein, W. B., 1962, The concept of entropy in landscape evolution: U.S. Geological Survey Professional Paper 500-A, 20 p.

Leopold, L. B., and Maddock, T., 1953, The hydraulic geometry of stream channels and some physiographic implications: U.S. Geological Survey Professional Paper 252, 57 p.

Leopold, L. B., and Wolman, M. G., 1957, River channel patterns: Braided, meandering, and straight: U.S. Geological Survey Professional Paper 282-B, p. 39–85.

Ljungner, C., 1930, Spaltentektonik und morphologie der schwedischen Skagerrak-Küste: Geological Institute of Upsala Bulletin, v. 21, p. 1–478.

Mackin, J. H., 1948, Concept of the graded river: Geological Society of America Bulletin, v. 59, p. 463–512.

Melton, M. A., 1957, An analysis of the relations among elements of climate, surface properties and geomorphology: Office of Naval Research Project NR 389-042, Technical Report 11, Department of Geology, Columbia University, New York, 102 p.

——1958a, Geometric properties of mature drainage systems and their representation in an E_4 phase space: Journal of Geology, v. 66, p. 35–54.

——1958b, Correlation structure of morphometric properties of drainage systems and their controlling agents: Journal of Geology, v. 66, p. 442–460.

Mendenhall, W. C., 1920, Memorial of Grove Karl Gilbert: Geological Society of America Bulletin, v. 31, p. 26–64.

Ollier, C. D., 1968, Open systems and dynamic equilibrium in geomorphology: Australian Geographical Studies, v. 6, p. 167–170.

Philippson, A., 1886, Studien über Wasserscheiden: Leipzig, Verhandlungen Erdkunde, 163 p.

Powell, J. W., 1875, Exploration of the Colorado River of the West: Washington, D.C., U.S. Government Printing Office, 291 p.

——1876, Report on the geology of the eastern portion of the Uinta Mountains and a region of the country adjacent thereto: Washington, D.C., U.S. Geographical and Geological Survey of the Territories, 218 p.

Pyne, S., 1975, The mind of Grove Karl Gilbert: 6th Annual Geomorphology Symposium, Binghamton, N.Y., Proceedings, p. 277–298.

——1976, Grove Karl Gilbert, A biography of American geology [Ph.D. dissert.]: Austin, University of Texas, 635 p.

Rayner, J. N., 1972, Conservation, equilibrium, and feedback applied to atmospheric and fluvial processes: Association of American Geographers, Committee on College Geography, Resource Paper no. 15, 23 p.

Scheidegger, A. E., 1964, Some implications of statistical mechanics in geomorphology: International Association of Scientific Hydrology Bulletin, v. 9, p. 12–16.

——1967, A complete thermodynamic analogy for landscape evolution: International Association of Scientific Hydrology Bulletin, v. 12, p. 57–62.

Scheidegger, A. E., and Langbein, W. B., 1966, Steady state in the stochastic theory of longitudinal river profile development: International Association of Scientific Hydrology Bulletin, v. 11, p. 43–49.

Schumm, S. A., 1956a, Evolution of drainage systems and slopes in badlands at Perth Amboy, New Jersey: Geological Society of America Bulletin, v. 67, p. 597–646.

——1956b, The role of creep and rainwash on the retreat ofbadland slopes: American Journal of Science, v. 254, p. 693–706.

——1963, The disparity between present rates of denudation and orogeny: U.S. Geological Survey Professional Paper 454-H, 13 p.

——1973, Geomorphic thresholds and complex response of drainage systems: 4th Annual Geomorphology Symposium, Proceedings, Binghamton, N.Y., p. 299–310.

——1975, Episodic erosion, A modification of the geomorphic cycle: 6th Annual Geomorphology Symposium, Proceedings, Binghamton, N.Y., p. 70–85.

——1977, The fluvial system: New York, John Wiley & Sons, Inc., 338 p.

Schumm, S. A., and Khan, H. R., 1972, Experimental study of channel patterns: Geological Society of America Bulletin, v. 83, p. 1755–1770.

Schumm, S. A., and Lichty, R. W., 1965, Time, space, and causality in geomorphology: American Journal of Science, v. 263, p. 110–119.

Smalley, I. J., and Vita-Finzi, C., 1969, The concept of "system" in the earth sciences, particularly geomorphology: Geological Society of America Bulletin, v. 80, p. 1591–1594.

Smith, A., 1832, On the water courses and the alluvial and rock formations of the Connecticut River valley: American Journal of Science, 1st ser., v. 22, p. 205–231.

Strahler, A. N., 1950, Equilibrium theory of erosional slopes, approached by frequency distribution analysis: American Journal of Science, v. 248, p. 673–696, 800—814.

——1952, Dynamic basis of geomorphology: Geological Society of America Bulletin, v. 63, p. 923–938.

Surell, A., 1841, Etude sur les torrents des Hautes-Alpes: Paris, Carilian-Goeury and V. Dalmont, 283 p.

Taylor, T. J., 1851, An inquiry into the operation of running streams and tidal waters: London, Longman, Brown, Green and Longmans, 119 p.

Trudgill, S. T., 1977, Soil and vegetation systems: Oxford University Press, 224 p.

Woldenberg, M. J., 1966, Horton's laws justified in terms of allometric growth and steady state in open systems: Geological Society of America Bulletin, v. 77, p. 431–434.

——1969, Spatial order in fluvial systems: Horton's laws derived from mixed hexagonal hierarchies of drainage basin areas: Geological Society of America Bulletin, v. 80, p. 97–112.

Wolman, M. G., 1955, The natural channel of Brandywine Creek, Pennsylvania: U.S. Geological Survey Professional Paper 271, 56 p.

Worssam, B. C., 1973, A new look at river capture and the denudation history of the Weald: London, Institude of Geological Sciences, 73/17, 21 p.

Yang, C. T., 1971, Potential energy and stream morphology: Water Resources Research, v. 7, p. 311–322.

MANUSCRIPT RECEIVED BY THE SOCIETY OCTOBER 9, 1979
MANUSCRIPT ACCEPTED MAY 20, 1980

Printed in U.S.A.

Geological Society of America
Special Paper 183
1980

Analogies in G. K. Gilbert's philosophy of science

DAVID B. KITTS
School of Geology and Geophysics and Department of the History of Science, University of Oklahoma, Norman, Oklahoma 73019

ABSTRACT

An examination of G. K. Gilbert's methodological works leaves some doubt as to the role he meant to assign to analogies in the quest for geological knowledge. The drawing of analogies can have only limited utility in a historical discipline, such as geology, in which it is supposed that each phenomenon encountered is assignable, at least in principle, to a kind of phenomenon identified by an extant physical theory.

INTRODUCTION

This volume stands as testimony to G. K. Gilbert's place in American geology. Successive generations have found in him the quintessential geologist, not simply to be revered as a towering figure in the history of our science, but to be emulated in the day-to-day practice of our discipline. His Henry Mountains report (Gilbert, 1877) is perhaps the most admired piece of investigation and writing in all of American geology. Gilbert was, moreover, unusual among distinguished geologists in that he left a small body of methodological work in which he attempted to stand outside geology and give a general account of how he and other geologists proceed in their search for knowledge. Contemporary American geologists seem to agree that no one has come closer than Gilbert to giving an adequate account of geological methodology. In this paper I shall examine some of Gilbert's methodological views, especially to determine just what role he meant to assign to analogies in the quest for geological knowledge, and I shall briefly consider the relevance of Gilbert's treatment of analogies to the practice of contemporary geology.

I shall begin by making three distinctions upon which, I believe, an understanding of any theory of geological inference must depend. They are the distinctions between theoretical and historical science, between the discovery and the justification of our knowledge, and between discourse about the world and discourse about discourse about the world. The distinction between historical science and theoretical science reflects a fundamental property of our natural language. We refer to particular events in singular statements. We locate them, describe them, and sometimes give them proper names. On the other hand, we formulate general and universal statements that refer to groups of things, sometimes as mere contingent collections of entities that share some accidental characteristic, and sometimes as kinds of things all of which have

some theoretically significant or essential property in common. History is concerned with particular events, with the activities of individual men, or with changes in bodies of rock conveyed in singular descriptive statements. But we cannot describe individual things except in terms of general predicates. To describe something is to assign to it properties that it shares with other things. Historical events, even as we attempt to establish their uniqueness, must always be seen as members of classes of events. And, moreover, descriptions of events that are separated from us in time dangle, often precariously, at the end of predictive and retrodictive inferences which must be justified by invoking general statements pointing to the association of events of certain kinds. And finally, these general statements, or natural laws as they are sometimes called, must themselves be justified by descriptions of events that are purported to be instances of them.

I have claimed (see especially Kitts, 1977) that geology is the paradigmatic historical science. If I am right, we should expect to find that geologists are primarily concerned with the derivation and testing of singular statements, and that when they formulate general principles it is with the view of employing them as instruments of historical inference. I maintain that an examination of the geological literature will strongly support this view. Theoretical scientists, on the other hand, see as their primary task the derivation and testing of general principles, especially those highly organized bodies of abstract statements called theories. But theories are no more selfjustifying than are singular statements. Theories must be tested against singular statements which are, according to them, relevant. My contention that geology is a historical science does not rest upon the conviction that geologists are not knowledgeable about scientific theories, nor that they do not sometimes test theories, nor even that they may not formulate theories upon rare occasions. It most emphatically does not depend upon the view that everyone who is called a physicist is engaged in theory construction. The distinction between historical science and theoretical science lies primarily in what those who practice the two disciplines see as their goal. For historical scientists, singular descriptive statements are ends, and theories are means to those ends. For theoretical scientists, theories are ends, and singular descriptive statements are means to those ends. There is another significant feature of geological methodology related to, but by no means following from, its historical character. Virtually without exception geologists are committed to the view that for the purposes of geological inference the most widely held physical theories of their day are not only inviolable but also are sufficient.

The distinction between the discovery and justification of knowledge is familiar, both in everyday discourse and in science, but it is one that geologists sometimes fail to take into account in their discussions of methodological issues. If we are to make any progress toward an understanding of the role of analogies in geological inference, we must plainly mark the difference between the question of how we come to arrive at some hypothesis and the question of how we are prepared to test it, or support it, or justify it.

Finally, it must be noted that there is a difference between *geology*, that body of statements which has as its object the Earth, and *metageology*, that body of statements which has as its object *statements* about the Earth. It was a failure to recognize this difference that led Baker and Pyne (1978, p. 98) to say, "The enigma of the reverence for Gilbert by modern geologists versus Gilbert's dissociation with the largely historical science of his contemporaries has even led to one recent study (Kitts, 1973) that mistakenly attempted to describe Gilbert's science as 'primary historical inference.' " I had not even attempted to describe Gilbert's science. In the paper to which Baker and Pyne allude, I confined my remarks to what Gilbert said about the science of geology and paid no heed to what he said about the Earth's crust. Let us now proceed to examine some of Gilbert's remarks *about* geology and especially those concerning the role of analogies in geological inference.

ANALOGIES IN GILBERT'S PHILOSOPHY OF SCIENCE

Early in his most famous paper on scientific methods, Gilbert (1886, p. 286) said,

It is the province of research to discover the antecedents of phenomena. This is done by the aid of hypothesis. A phenomenon having been observed, or a group of phenomena having been established by empiric classification, the investigator invents an hypothesis in explanation. He then devises and applies a test of the validity of the hypothesis. If it does not stand the test he discards it and invents a new one. If it survives the test, he proceeds at once to devise a second. And thus he continues until he finds an hypothesis that remains unscathed after all the tests his imagination can suggest.

In the same paper he attended to the question of the origin of hypotheses in an often quoted passage:

Given a phenomenon A, whose antecedents we seek. First we ransack the memory for some different phenomenon, B. which has one or more features in common with A, and whose antecedent we know. Then we pass by analogy from the antecedent of B, to the hypothetical antecedent of A, solving the analogic proportion—as B is to A, so is the antecedent of B to the antecedent of A. [Gilbert 1886, p. 287]

According to Gilbert, a plexus of antecedent and consequent relations pervades nature. Scientific understanding does not, in his view, consist of coming to see the events of that plexus, and of any plexus that ever has been, ever will be, or ever could be, as instances of some comprehensive theory. Scientific research is directed at revealing the antecedent and consequent relationships among those particular events which in fact comprise the history of the Earth.

Gilbert made a clear distinction between discovery and justification. The drawing of analogies is wholly concerned with the "discovery," or the "invention," or the "generation" of hypotheses. Analogies play no role in the justification or, to use Gilbert's term,

in the "testing" of hypotheses. If a phenomenon A has antecedents we know and has one or more features in common with a phenomenon B whose antecedents we seek, this may suggest to us, and *only* suggest to us, that B has the same kinds of antecedents as A. The justification for the hypothesis is to be sought in a test which is completely independent of the analogy. For Gilbert the problems of *argument* from analogy and of the *logic* of analogy, which have occupied so much of the attention of philosophers, did not arise. Argument and logic reside in test procedures, not in the discovery of hypotheses.

Gilbert (1896, p. 2) suggested that hypotheses about the antecedents of phenomena are always generated out of analogies, but it is plain that we often discover the antecedents of phenomena by a process in which drawing analogies plays no role. Someone might, for example, be able to specify correctly the antecedents of a particular eclipse of the sun while denying that drawing an analogy between the eclipse and some other event had played any role whatever in the discovery of the antecedents. Different events that, according to some widely accepted theory, have certain essential properties in common are not simply analogous, they are of the same *kind*. We need not seek the antecedents of an eclipse of the sun in an analogy because we know a theory according to which all eclipses of the sun are of the same kind and according to which, moreover, all eclipses of the sun have the same kinds of antecedents. The situation is not much different in the case of a geologist who encounters a vesicular basalt. His conclusion about the antecedent conditions which explain the origin of *that* basalt may not have been discovered by ransacking his memory for some other phenomenon. It may have fallen virtually unbidden out of a general explanatory theory of the origin of vesicular basalts, a theory which a geologist might know and be able to apply in the absence of any memory of a previous encounter with a basalt.

You may object that Gilbert was not concerned with a method for discovering the antecedents of already familiar kinds of phenomena such as eclipses of the sun and the formation of vesicular basalts. He might even have denied that the antecedents of such familiar phenomena were in need of *discovery*. Gilbert was concerned with the method by which we develop hypotheses about the antecedents of *unfamiliar* phenomena such as that peculiar depression in northern Arizona (Gilbert, 1896). But can we so clearly distinguish between familiar phenomena and unfamiliar phenomena? There seems to be every degree of familiarity, from phenomena so familiar as a falling rock to those so unfamiliar as the impact of a meteor. But it is a striking feature of contemporary geology that geologists never encounter phenomena which they regard as so unfamiliar as to be in need of explanation in terms of an as yet unformulated comprehensive theory. Although a specimen of vesicular basalt is in some obvious sense more familiar than Meteor Crater, the two are alike in that in both cases the search for antecedents will be guided by our general knowledge of the world, knowledge that finds its most explicit and formal expression in geological principles and physical theories. Whatever the outcome of our search for antecedents, it will turn out in such a way as not to violate these principles and theories. Well-known and widely accepted geological principles and physical theories provide not only our only means of justifying hypotheses about the antecedents of phenomena, but our principal source of such hypotheses.

In suggesting that drawing analogies provides the only means of generating hypotheses, Gilbert seems to have relegated our systematized general knowledge to the role of justifying and testing hypotheses. There is some evidence, however, that in Gilbert's treatment of geological method is a recognition of the crucial role played by geological principles and comprehensive physical theories in the generation of hypotheses. There is, after all, a certain ambiguity in the claim that we ransack our memories for *a* phenomenon. According to Aristotle, things can be one in number or one in kind. We use "phenomenon" in both senses. Thus, we may speak of the phenomenon of the American Civil War and the phenomenon of the retrogression of the planets. It is not clear in which sense we are to understand Gilbert's use of the word. In his suggestion that certain structures in local mines were the source of the hypothesis that Meteor Crater was formed by an explosion (Gilbert, 1896, p. 3), he seems to invoke memory as a means of calling up a mental picture of some particular past phenomenon. At times, on the other hand, it seems that ransacking our memory is just Gilbert's way of talking about the process by which we are reminded of our general knowledge. Consider, for example, the following passage (Gilbert, 1886, p. 287) which closely follows his account of the analogic proportion:

The consequential relations of nature are infinite in variety, and he who is acquainted with the largest number has the broadest base for the analogic suggestion of hypotheses. It is true that a store of scientific knowledge cannot take the place of mental strength and training, i.e., of functional ability inherited and acquired, but it is nevertheless a pre-requisite of fertility of hypothesis.

We can all agree that the body of scientific knowledge is not exhausted by a catalogue of phenomena. It consists, in very significant part, of principles, laws, and theories that permit us to see phenomena as instances of theoretically significant categories. Gilbert seems to grant that general knowledge may serve as a source of hypotheses. Where general knowledge provides the source of hypotheses, memory does not serve to produce a mental picture of some phenomenon previously encountered, but simply to call up our knowledge *that* something is the case—*that*, for example, force is equal to the product of mass and acceleration.

There is yet another possible interpretation of Gilbert's views on the role of analogy in geology. Perhaps he did not intend to instruct geologists as to how to proceed in the search for hypotheses, but rather to give them an account of what goes on in their minds, wholly unnoticed, when they make a discovery. Gilluly is apparently inclined toward this view (see Gilluly, 1963, p. 222). In the famous passage on the role of analogy in discovery quoted above, Gilbert could be giving us directions to "ransack our memory," and "solving the analogic proportion" sounds like the sort of thing that we could learn to do quite consciously. But consider the following treatment of discovery written 10 yr later (Gilbert, 1896, p. 2):

The mental process by which hypotheses are suggested is obscure. Ordinarily they flash into consciousness without premonition, and it would be easy to ascribe them to a mysterious intuition or creative faculty; but this would contravene one of the broadest generalizations of modern psychology. Just as in the domain of matter nothing is created from nothing, just as in the domain of life there is no spontaneous generation, so in the domain of mind there are no ideas which do not owe their existence to antecedent ideas which stand in the relation of parent to child. It is only because our mental processes are largely conducted outside the field of consciousness that the lineage of ideas is difficult to trace.

To explain the origin of hypotheses I have a hypothesis to present,—not, indeed, as original, for it has been at least tacitly assumed by various writers on scientific method, but rather as worthy of more general attention and recognition. It is that hypotheses are always suggested through analogy.

There is a question as to just what role Gilbert meant to assign to analogies in the acquisition of geological knowledge. Recent discussions of Gilbert's methodology by Pyne (1978, 1979) and by Baker and Pyne (1978) have contributed nothing toward an answer to this question. Baker and Pyne (1978, p. 103) in their discussion of the Henry Mountains report (Gilbert, 1877) stated, "Having described the uplift in dynamic terms, he described the erosion of streams according to the concept of energy. The stream, like the uplift process, was analogous to a machine that performed work according to the laws of thermodynamics." We must not be misled into supposing that Baker and Pyne are simply paraphrasing Gilbert's account of his work in the Henry Mountains. Nowhere in his report did Gilbert say that the stream is analogous to a machine that performs work according to the laws of thermodynamics. To analyze this famous specimen of geological investigation as an instance of analogical reasoning is not only historically unjustified, it unnecessarily complicates what appears to be a perfectly straightforward and conventional case of geological inference. If there is any hypothesis involved in applying thermodynamics to geology, it is one to the effect that certain principles of physics that purport to apply to one domain apply to another. But geologists do not recognize separate domains whose comprehension by a physical theory must be hypothesized and tested. For Gilbert, thermodynamic theory, or rather his covertly formulated version of it, has as its domain any physical system. The rocks and streams of the Henry Mountains comprise a physical system. He introduces energy into his discussion of land sculpture without any preamble whatever. The notion that we can apply the concept of energy to geological phenomena is no more in need of explanation in terms of some theory of discovery than is the notion that we can apply the concept of density to rocks. And to suggest that the application of thermodynamics to a stream depends upon an analogy between streams and machines is like suggesting that the application of classical mechanics to the Earth's crust depends upon an analogy between the behavior of rocks and the orbital motion of the planets.

AN ANALOGY IN GILBERT'S GEOLOGY

Neither Gilbert nor Baker and Pyne consider the question, so interesting to physicists and philosophers of science, of how analogies might function in the formulation of comprehensive physical theories such as thermodynamics which geologists presuppose. It has been suggested that new physical theories which permit us to recognize new kinds of events are often discovered with the aid of analogies. But what role can analogies play in a radically historical discipline such as geology, in which the assumption is made that every event encountered is a kind of event already identified by some extant theory? This assumption does not entail that we will always know into which kind, identified by which theory, the event falls. But geologists, including Gilbert, have not supposed that any event that they encounter is novel, which is to say, that it requires explanation in terms of some as yet undiscovered theory.

In view of these considerations, it seems best to consider a hypothesis about the antecedents of a phenomenon in which another phenomenon is invoked as a hypothesis to the effect that the two events fall into the same theoretically significant category. Gilbert's example which points to the explosion structures in mines as suggestive of the origin of Meteor Crater is in accord with this view. One plausible way of interpreting Gilbert's position on the role of analogies is that for him phenomena which are only hypothesized to be of the same kind are analogous. But the key to an understanding of Baker and Pyne's view of analogies, and probably also of Gilbert's lies in a recognition of the fact that they suppose that two events may be at once of the same kind and analogous. Kinds of events are, after all, not exactly alike, but only alike in certain essential respects. Suppose that we know enough about an event A to identify it as of a certain kind under a theory T, and, in addition, know certain properties of A that are nonessential according to T. Suppose further that we are interested in another event B about which we know only enough to recognize that it is of the same kind as A. Someone might hypothesize that in addition to the essential properties of B, which identified it as of the same kind as A, B shared with A some of those nonessential properties known to occur in A. Because the hypothesis is not to be justified theoretically or empirically, it might be supposed that it had been suggested by an analogy, particularly because it seems crucially to involve two events. The projection of the properties of one event to the other seems to be wholly gratuitous. It would be not quite so gratuitous and more analogical if the projection were based upon the view that, because each event is itself a complex causal nexus, the more properties two events have in common the more likely they are to have other properties in common. But so long as we did not seek to justify or test, but only to discover or invent a hypothesis on the basis of the projection, the question of its gratuitousness would not matter and, in fact, need not even arise.

Whatever the precise character of the procedure I have described above, I think that it can be shown to play a significant role in geology. That role is revealed more plainly in Gilbert's account of his geological investigations than it is in his explicit treatment of scientific methodology. Let us consider, as a case in point, his work "The Transportation of Debris by Running Water," published in 1914. Pyne (1978, p. 418) said of this justly famous specimen of geology, "When in his later years he tried to extrapolate from laboratory flumes to natural streams by recommending that flume data might apply to streams which were geometrically similar, he was establishing such an 'analogic proportion'." Gilbert does not once mention analogies in this report, but he does tell us a good deal about his reasons for undertaking the flume experiments. In the preface of his paper, he stated (Gilbert, 1914, p. 9),

Thirty-five years ago the writer made a study of the work of streams in shaping the face of the land. The study included a qualitative and partly deductive investigation of the laws of transportation of debris by running water; and the limitations of such methods inspired a desire for quantitative data, such as could be obtained only by experimentation with determinate conditions.

There was no need for Gilbert to solve an analogic proportion nor to give any account whatever of how he came to suppose that he could learn something about natural streams from the study of flumes. Flumes and streams are alike in their determinate conditions. Gilbert did not have to perform an experiment to demonstrate this. What counts as a determinate condition has already been dictated by his general knowledge of the physical world. He undertook his flume experiments because the determinate conditions identified by the theories which he presupposed could be dealt with more readily in a controlled environment than the *same* determinate conditions could be got at in the field.

I have insisted that the explication of geological methodology rests crucially upon the distinction between events that are of the same kind and events that are analogous. But the distinction which is so plain upon abstract analysis appears not to be so distinct in particular instances. Geologists, being strongly inclined to see events as significant in their own right rather than as mere instances of some theory, are more concerned with the differences among events than are physicists. Events that are simply regarded as of the same kind by a physicist because they are comprehended by the same theory may, from the perspective of one preoccupied with history, be only "analogous." It is difficult to imagine that a physicist, *qua* physicist, would regard water running in a flume and water running in a stream as anything but events of the same kind. Yet Pyne (1978) suggested that Gilbert saw them as analogous. Leaving aside the already mentioned fact that Gilbert did not tell us that he saw them as analogous, we shall pursue the question of what any other geologist might mean if he were to claim that two events so obviously comprehended by the same physical theories were analogous. He might mean that although it is to be admitted that water running in any channel whatever comprises a theoretically significant kind, at the level of resolution required for historical analysis there are many *different kinds* of streams and rivers, and that, indeed, the final goal of historical analysis might be to see each stream as unique, which is to say, as a *one of a kind* phenomenon.

Now a physicist might claim that *in principle* all these different streams could be described in terms of the universal predicates of the theory which, it is agreed, comprehends them. But geologists will rightly maintain that it is impossible to give a satisfactory, let alone a complete, account of a real stream wholly in terms of such predicates. It is no comfort to be assured that this impossibility is not theoretically necessary, but is only contingent upon the vast complexity of real systems. And so the geologist, albeit guided by an appropriate theory, attempts to cope experimentally with the concrete cases, to contrive situations which in their complexity fall somewhere between the highly idealized limiting cases of physical theory and the overwhelming and unmanageable richness of geological phenomena. A geological experiment is not, like an experiment in physics, directed toward the testing of a theory. The theory is, for the sake of the experiment, presupposed. A geological experiment is, as it were, pointed in the other direction, toward the illumination of concrete geological events. If anything is being *tested* in a geological experiment, it is the "fit" between abstract physical theories and those concrete geological events. Contemporary geologists often speak of these experiments as "physical models," thereby pointing to the attempt not to reproduce geological phenomena is the laboratory but rather, through the process of simplification, only to "model" them. But there is a difficulty, almost a paradox, that lies at the heart of geological modeling and experimenting. Almost always, it seems, if an experiment does justice to the richness of a geological phenomenon, it fails to achieve the purpose of the experiment, which is to contrive a situation in which the appropriate initial and boundary conditions can be recognized and quantified. There is a limit to how simple a phenomenon can be and still be a phenomenon of geological significance. We can sense Gilbert's frustration in the face of this fact when he said (1914, p. 240):

It was the primary purpose of the Berkeley investigation to determine for rivers the relation which the load swept along the bed bears to the more important factors of control. As a means to that end it was proposed to study the mode of propulsion and learn empirically the laws connecting its output with each factor of control taken separately. The review of results in the present chapter shows that the primary purpose was not accomplished. In the direction of the secondary purpose much more was achieved, and a body of definite information is contributed to the general subject of stream work. A valuable outcome is the knowledge that the output in tractional load is related to the controlling conditions in a highly complex manner, the law of control for each condition being qualified by all other conditions.

Despite Gilbert's apparent disappointment, the following passage reveals his willingness to project cautiously what has been learned from the flume experiments to natural streams (Gilbert, 1914, p. 240):

It is thought that the laboratory formulas may be applied to natural streams which are geometrically similar to the laboratory streams—that is, to streams having the same slopes and form ratios and carrying debris of proportionate size. The class of streams to which the formulas apply by reason of similarity is necessarily restricted, being characterized in the main by high slopes and coarse debris. It can include few large streams.

Gilbert would apply his empirically derived formulas only to streams that are almost exactly like the flume. Because all of the determinate factors are so complexly interrelated, it is certainly plausible to suppose that this knowledge he has about a flume might be applied to a stream that he knows to be like it in almost every other way. But Gilbert made no claim that such a projection can be justified. A "chasm" separates the flume from the stream that can only be bridged by further work. He died a few years after his study was published, but he knew just how the work should proceed. He said (1914, p. 240):

It is possible that the chasm between the laboratory and the river may be bridged only by an adequate theory, the work of the hydromechanist. It is possible also that it may be practically bridged by experiments which are more synthetic than ours, such experiments as may be made in the model rivers of certain German laboratories.

The justification for the projection from experiment to river must come either from a theory that compels us to suppose that the two are alike in all relevant respects, or from empirical studies that permit us to identify every relevant property of the stream with a property of the model. I have insisted that geologists suppose that all geological phenomena are explicable in terms of the most comprehensive and widely accepted theories of their day. I have not claimed, nor do I wish to claim, that such explanations are always achieved nor that a failure to achieve them deters a geologist from continuing his investigations. In the example at hand, so long as experimental studies lead to the recognition of events and structures in the field and so long as field studies suggest alterations in the model, confidence in the relevance of the model to the geological phenomena will increase whether or not theoretical justification for such relevance is forthcoming. It would be merely perverse to deny that some understanding of the phenomena in question had been achieved in the process, even in the absence of that deeper understanding which comes from demonstrated comprehension by some physical theory.

If there is a significant role for analogies in geology, it lies in physical modeling involving two phenomena that are known to have some properties in common and that are hypothesized, without immediate theoretical or empirical justification, to have other properties in common. The fate of the analogies that arise in geological modeling is not, as is the case with analogies in physics, to be discarded when they have played some role in discovery or to continue to be invoked as heuristic devices. Rather, they have their analogic status revoked as they are identified as to kind.

SUMMARY AND CONCLUSIONS

I have attempted to show that analogies play a limited role in the acquisition of geological knowledge. The reason that they do is clear. Analogies are useful in cases where there is some uncertainty about the theoretical structure of the world with which we have to deal. Thus, analogies have been thought to have their most important application in the invention, or the discovery, of new theories. Geologists accept, simply as a methodological rule or sometimes on deeper metaphysical grounds, that the theoretical apparatus that they presuppose is a sufficient instrument for coping with the phenomena they will encounter. Each geological event is already assumed to be assignable, at least in principle, to some kind of event defined by an available theory. Contemporary geology accepts the rational restraint provided by the most comprehensive physical theories. The task of bridging the enormous gap between the idealized accounts of the world contained in physical theories and the richness of the world of concrete events is an extremely difficult one. For this reason we reserve our greatest acclaim for those geologists who, like Gilbert, find new ways to apply extant theories in geological inferences.

I do not deny that some geological hypotheses are discovered in much the way Gilbert has suggested. If I were to see a hard, red substance in contact with a basalt and grading into a loose clay, I might be reminded of a brick and of what I know about the antecedents of bricks. I cannot imagine that I would, under these circumstances, do anything so deliberate as to ransack my memory, nor so formal as to solve an analogic proportion. It is for this reason that I prefer to think (for I plainly have a choice) that Gilbert was speaking very loosely when he suggested that such conscious acts attend the generation of hypotheses and that he really meant to point to some obscure process by which hypotheses "flash into consciousness without premonition."

It is surprising that the brief remarks which Gilbert devoted to analogies should have assumed so much importance for the geologists who followed him. Analogies did not play a central role in Gilbert's philosophy of science. His principal methodological point was that the advance of scientific knowledge depended upon the generation of hypotheses which could be subjected to rigorous test. He presented the *hypothesis* that hypotheses are suggested through analogy. Nowhere did he claim to have subjected *that* hypothesis to a test.

It must be noted in closing that Gilbert (see especially Gilbert, 1896) regarded the practice of geology as difficult to describe, let alone prescribe. He believed that we must learn to do geology by emulating those who have done it well. Gilbert's stature in American geology has never rested upon the few remarks he made about analogies nor even upon the entire body of his methodological work. It has rested upon the high quality of a large body of geological work which we can emulate with profit.

REFERENCES CITED

Baker, V. R., and Pyne, S. J. 1978, G. K. Gilbert and modern geomorphology: American Journal of Science, v. 278, p. 97–123.

Gilbert, G. K., 1877, Report on the geology of the Henry Mountains: U.S. Geological Survey of the Rocky Mountain Region, 160 p.

——1886, The inculcation of scientific method by example: American Journal of Science, 3d ser., v. 31, p. 284–299.

——1896, The origin of hypotheses: Science, n.s. 3, v. 53, p. 1–13.

——1914, The transportation of debris by running water: U.S. Geological Survey Professional Paper 86, 263 p.

Gilluly, J., 1963, The scientific philosophy of G. K. Gilbert, *in* Albritton, C. C., ed., The fabric of geology: Reading, Massachusetts, Addison-Wesley, p. 218–224.

Kitts, D. B., 1973, Grove Karl Gilbert and the concept of "hypothesis" in late nineteenth-century geology, *in* Giere, R. N., and Westfall, R. S., eds., Foundations of scientific method: The nineteenth century: Bloomington, Indiana University Press, p. 259–274.

——1977, The structure of geology: Dallas, Southern Methodist University Press, 180 p.

Pyne, S. J., 1978, Methodologies of geology. G. K. Gilbert and T. C. Chamberlin: Isis, v. 69, p. 413–424.

——1979, Certain allied problems in mechanics: Grove Karl Gilbert at the Henry Mountains, *in* Schneer, C. J., ed., Two hundred years of geology in America: Hanover, New Hampshire, University Press of New England, p. 225–238.

MANUSCRIPT RECEIVED BY THE SOCIETY DECEMBER 26, 1979
MANUSCRIPT ACCEPTED MAY 20, 1980